YES New SAT Math

WORKBOOK

YES New SAT Math

Calculation Section (37 problems) 55 min

Multiple Choice	30 questions	1 point	75%
Student-Produced Response	6 questions	1 point	15%
Extended Thinking	1 question	4 points	10%

Content Categories

Heart of Algebra: 13 questions / Problem Solving and Data Analysis: 14 questions

Passport to Advanced Algebra: 7 questions / Additional Topic in Math: 3 questions

No Calculation Section (20 problems) 25 min

Multiple Choice	15 questions	1 point	75%
Student-Produced Response	5 questions	1 point	25%

Content Categories

Heart of Algebra: 8 questions / Problem Solving and Data Analysis: 0 questions

Passport to Advanced Algebra: 9 questions / Additional Topic in Math: 3 questions

Total Content Categories

Heart of Algebra: 21 questions 35% / **Problem Solving and Data Analysis**: 14 questions 28%

Passport to Advanced Algebra: 18 questions 27% / **Additional Topic in Math**: 6 questions 10%

Heart of Algebra: 21 questions 35%

Interpret linear equation or linear inequalities in one or two variable. Solve systems of equations

Problem Solving and Data Analysis: 14 questions 28%

Proportional / Relationships / Percentages / Data Interpretation and Synthesis

Passport to Advanced Algebra: 18 questions 27%

Quadratic or Exponential Functions / Radicals and Rational Expression / Solve Quadratic

Equation / Polynomials / Radical

Additional Topic in Math: 6 questions 10%

New SAT Math Problems from College Board

1. Aaron is staying at a hotel that charges $99.95 per night plus tax for a room. A tax of 8% is applied to the room rate, and an additional one-time untaxed fee of $5.00 is charged by the hotel. Which of the following represents Aaron's total charge, in dollars, for staying x nights? (Calculator)

(A) $(99.95 + 0.08x) + 5$ (B) $1.08(99.95x) + 5$ (C) $1.08(99.95x + 5)$

(D) $1.08(99.95 + 5)x$

2. The gas mileage for Peter's car is 21 miles per gallon when the car travels at an average of 50 miles per hour. The car's gas tank has 17 gallons of gas at the beginning of a trip. If Peter's car travels at an average speed of 50 miles per hour, which of the following functions f models the number of gallons of gas remaining in the tank t hours after the trip begins? (Calculator)

(A) $f(t) = 17 - \dfrac{21}{50t}$ (B) $f(t) = 17 - \dfrac{50t}{21}$ (C) $f(t) = \dfrac{17 - 21t}{50}$

(D) $f(t) = \dfrac{17 - 50t}{21}$

3. If $-\dfrac{9}{5} < -3t + 1 < -\dfrac{7}{4}$, what is one possible value of $9t - 3$? (Calculator)

4. In the equation below, what is the value of k? (No Calculator)

$$\frac{5(k+2) - 7}{6} = \frac{13 - (4-k)}{9}$$

(A) $\dfrac{9}{17}$ (B) $\dfrac{9}{13}$ (C) $\dfrac{33}{17}$ (D) $\dfrac{33}{13}$

5. Based on the system of equations below, what is the value of the product xy?

$$\begin{cases} 4x - y = 3y + 7 \\ x + 8y = 4 \end{cases}$$

(A) $-\dfrac{3}{2}$ (B) $\dfrac{1}{4}$ (C) $\dfrac{1}{2}$ (D) $\dfrac{11}{9}$ (No Calculator)

6. If $\dfrac{1}{2}x + \dfrac{1}{3}y = 4$, what is the value of $3x + 2y$?   (No Calculator)

8. The toll rates for crossing a bridge are \$6.50 for a car and \$10 for a truck. During a two-hour period, a total of 187 cars and trucks crossed the bridge, and the total collected in tolls was \$1,338. Solving which of the following systems of equations yields the number of cars, x, and the number of trucks, y, that crossed the bridge during the two hours?   (Calculator)

(A) $x + y = 1338$
$6.5x + 10y = 187$

(B) $x + y = 187$
$6.5x + 10y = 1338$

(C) $x + y = 187$
$6.5x + 10y = \dfrac{1338}{2}$

(D) $x + y = 187$
$6.5x + 10y = 1338 \times 2$

9. In the system of linear equations below, a is a constant. If the system has no solution, what is the value of a? (No Calculator)

$$\frac{1}{2}x - \frac{1}{4}y = 5$$
$$ax - 3y = 20$$

(A) $\dfrac{1}{2}$ (B) 2   (C) 6   (D) 12

10. When a scientist dives in salt water to a depth of 9 feet below the surface, the pressure due to the atmosphere and surrounding water is 18.7 pounds per square inch. As the scientist descends, the pressure increases linearly. At a depth of 14 feet, the pressure is 20.9 pounds per square inch. If the pressure increases at a constant rate as the scientist's depth below the surface increases, which of the following linear models best describes the pressure p in pounds per square inch at a depth of d feet below the surface?   (Calculator)

(A) $p = 0.44d + 0.77$ (B) $p = 0.44d + 14.74$ (C) $p = 2.2d - 1.1$
(D) $p = 2.2d - 9.9$

11.   In a sequence, each term after the first is founded by adding the constant c to the preceding term. The 12th term in the sequence is 32, and the 18th term is 56. What is the 9th term in the sequence?

   (A) -16
   (B) -12
   (C) 20
   (D) 16

12.   The students in a certain physical education class are on either a basketball team or a tennis team, are on both these team, or not on either team. If 15 students are on the basketball team, 18 students are on the tennis team, 11 students are on the both teams, and 14 students are not any of these teams, how many students are in the class?

   (A) 28
   (B) 36
   (C) 38
   (D) 40

13. A typical image taken of the surface of Mars by a camera is 11.2 gigabits in size. A tracking station on Earth can receive data from the spacecraft at a data rate of 3 megabits per second for a maximum of 11 hours each day. If 1 gigabit equals 1,024 megabits, what is the maximum number of typical images that the tracking station could receive from the camera each day? (Calculator)

(A) 3   (B) 10   (C) 56   (D) 144

	Voted	Did Not Vote	No Response	Total
18-34 yrs old	30,329	23,211	9,468	63,008
35-54 yrs old	47,085	17,721	9,476	74,282
55-74 yrs old	43,075	10,092	6,831	59,998
74 & over old	12,459	3,508	1,827	17,794
Total	132,948	54,532	27,602	215,082

A survey was conducted among a randomly chosen sample of U.S. citizens aboutU.S. voter participation in the November 2012 presidential election. The table above displays a summary of the survey results. Reported Voting by Age (in thousands)

14. According to the table, for which age group did the greatest percentage of people report that they had voted? (Calculator)

(A) 18- to 34-year-olds   (B) 35- to 54-year-olds   (C) 55- to 74-year-olds

(D) People 75 years old and over

15. Of the 18- to 34-year-olds who reported voting, 500 people were selected at random to do a follow-up survey where they were asked which candidate they voted for. There were 287 people in this follow-up survey sample who said they voted for Candidate A, and the other 213 people voted for someone else. Using the data from both the follow-up survey and the initial survey, which of the following is most likely to be an accurate statement? (Calculator)

(A) About 123 million people 18 to 34 years old would report voting for Candidate A in the November 2012 presidential election.
(B) About 76 million people 18 to 34 years old would report voting for Candidate A in the November 2012 presidential election.
(C) About 36 million people 18 to 34 years old would report voting for Candidate A in the November 2012 presidential election.

(D) About 17 million people 18 to 34 years old would report voting for Candidate A in the November 2012 presidential election.

16. An international bank issues its Traveler credit cards worldwide. When a customer makes a purchase using a Traveler card in a currency different from the customer's home currency, the bank converts the purchase price at the daily foreign exchange rate and then charges a 4% fee on the converted cost.

Sara lives in the United States, but is on vacation in India. She used her Traveler card for a purchase that cost 602 rupees (Indian currency). The bank posted a charge of $9.88 to her account that included the 4% fee.   (Extended Thinking)

PART 1

What foreign exchange rate, in Indian rupees per one U.S. dollar, did the bank use for Sara's charge? Round your answer to the nearest whole number.

PART 2

A bank in India sells a prepaid credit card worth 7,500 rupees. Sara can buy the prepaid card using dollars at the daily exchange rate with no fee, but she will lose any money left unspent on the prepaid card. What is the least number of the 7,500 rupees on the prepaid card Sara must spend for the prepaid card to be cheaper than charging all her purchases on the Traveler card? Round your answer to the nearest whole number of rupees.

17. If $a^2 + 14a = 51$ and $a > 0$, what is the value of $a + 7$? (No Calculator)

18. The function f is defined by $f(x) = 2x^3 + 3x^2 + cx + 8$ where c is a constant. In the xy-plane, the graph of f intersects the x-axis at the three points $(-4, 0)$, $(\frac{1}{2}, 0)$ and $(p, 0)$. What is the value of c ? (Calculator)

(A) −18   (B) −2   (C) 2   (D) 10

19. The graph of $y=f(x)$ is shown above. If $f(3)=h$, which of the following could be the value of $f(h)$?

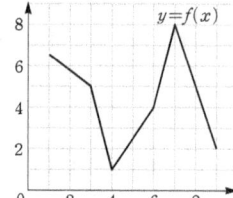

(A) 2
(B) 2.5
(C) 3
(D) 3.5

20. What is one possible solution to the equation $\dfrac{24}{x+1} - \dfrac{12}{x-1} = 1$? (No Calculator)

21. Anise needs to complete a printing job using both of the printers in her office. One of the printers is twice as fast as the other, and together the printers can complete the job in 5 hours. The equation below represents the situation described. Which of the following describes what the expression $\dfrac{1}{x}$ represents in this equation? (No Calculator)

$$\frac{1}{x} + \frac{2}{x} = \frac{1}{5}$$

(A) The time, in hours, that it takes the slower printer to complete the printing job alone
(B) The portion of the job that the slower printer would complete in one hour
(C) The portion of the job that the faster printer would complete in two hours
(D) The time, in hours, that it takes the slower printer to complete $\dfrac{1}{5}$ of the printing job

22. If (x,y) is a solution to the system of equations below, what is the value of x^2?

$$x^2 + y^2 = 153$$
$$y = -4x$$

(A) −51   (B) 3   (C) 9   (D) 144   (Calculator)

23. If the expression $\dfrac{4x^2}{2x-1}$ is written in the equivalent form $\dfrac{1}{2x-1}+A$, what is A in terms of x? (Calculator)

(A) $2x + 1$ (B) $2x - 1$ (C) $4x^2$ (D) $4x^2 - 1$

24. The figure on the right shows a metal hex nut with two regular hexagonal faces and a thickness of 1 cm. The length of each side of a hexagonal face is 2 cm. A hole with a diameter of 2 cm is drilled through the nut. The density of the metal is 7.9 grams per cubic cm. What is the mass of this nut, to the nearest gram? (Density is mass divided by volume.) (Calculator)

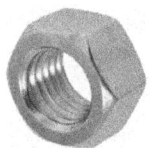

25. In the diagram on the right, each circle is divided into two equal areas and O is the center of the larger circle.
The area of the larger circle is 64π.
What is the total area of the shaded regions?

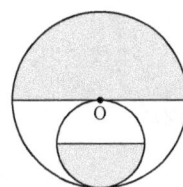

(A) 33π
(B) 34π
(C) 36π
(D) 40π

26. It is given that $\sin x = a$, where x is the radian measure of an angle and $\dfrac{\pi}{2} < x < \pi$. If $\sin w = -a.$, which of the following could be the value of w?

(A) $\pi - x$ (B) $x - \pi$ (C) $2\pi + x$ (D) $x - 2\pi$ (No Calulator)

YES New SAT Math Practice

1. At 1:00 p.m. a car leaves St. Louis for Chicago, traveling at a constant speed of 65 miles per hour. At 2:00 p.m. a truck leaves Chicago for St. Louis, traveling at a constant speed of 55 miles per hour. If it is a 305-mile drive between St. Louis and Chicago, at what time will the car and truck pass each other?

(A) 2:30 p.m.   (B) 3:00 p.m.   (C) 4:00 p.m.   (D) 4:30 p.m.

2. For all $x \neq 0$ and $y \neq 0$, $\dfrac{(2x^{-3}y^4)^3}{(4xy)^2} =$

(A) $\dfrac{y^{10}}{2}$ (B) $\dfrac{2y^{10}}{x^{10}}$ (C) $\dfrac{y^{10}}{2x^{11}}$ (D) $\dfrac{y^{11}}{2x^2}$

3. What is the sum of $(x+2y)^2$ and $(x-y)^2$?

(A) $2x^2 + 3y^2$ (B) $2x^2 + 5y^2$ (C) $2x^2 + 2xy + 5y^2$ (D) $4x^2 + y^2$

4. What are the factors of $12c^2 + cd - 6d^2$?

(A) $(4c+3d)(3c-2d)$ (B) $(4c-3d)(3c+2d)$ (C) $(6c+d)(2c-6d)$
(D) $6(2c-d)(c+d)$

5. How much larger than the sum of -1, -2, and 5 is the sum of the squares of these same integers?

(A) 18   (B) 20   (C) 22   (D) 28

6. Which of the following expresses the complete solution, for x, to the inequality shown? $3x-5 > 5x-9$

(A) $x > -\dfrac{7}{4}$ (B) $x > 2$ (C) $x < 2$ (D) $x > -2$

7. If $x^3+5=27$, then $x^3-5=$

(A) 12
(B) 17
(C) 39
(D) 144

8. The owner of a store displays a large jar of nickels and dimes and offers the value of the coins to the person who guesses how many dimes there are. If there are 1,130 coins, and they are worth $100, how many dimes are there?

(A) 130   (B) 260   (C) 870   (D) 970

9. In simplified form, $\dfrac{24x^4z^{-2}}{8x^{-3}z}=$

(A) $\dfrac{3x}{z}$ (B) $\dfrac{16x^2}{z}$ (C) $\dfrac{3x^7}{z^3}$ (D) $16xz$

10.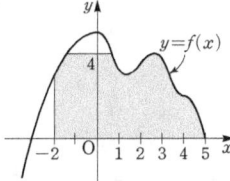

The shaded region in the figure above is bounded by x-axis, $x=-2$, $y=4$ and the graph $y=f(x)$. If the point (p, q) lies in the shaded region, which of the following must be true?

I. $p\leq4$ II. $p\leq q$ III. $q\leq f(p)$

(A) I only
(B) III only
(C) I and III only
(D) II and III only

11. If $y = \dfrac{2x+1}{x-5}$ and y is a real number, then

x CANNOT equal which of the following?

(A) 1
(B) 2
(C) 3
(D) 5

12. Tickets to a play cost $5 for adults and $2 for children. If 1,750 tickets were sold for a total of $7,100, how many children' tickets were sold?

(A) 550   (B) 650   (C) 1,100   (D) 1,200

13. Where are the points $(1, 2)$ and $(-1, 1)$ in relation to the line $3x + 4y = 7$?

(A) Both are on the line.

(B) One is on the line and the other is off the line.

(C) Both are above the line.

(D) One is above the line and the other is below the line.

14. If $f(x) = 2x + 7$, $g(x) = 3x - 5$, and $h(x) = 2x + 6$, then $h(x) + [f(x) \cdot g(x)] - 6 =$

(A) $7x + 2$ (B) $7x + 8$ (C) $6x^2 + 13x - 29$ (D) $6x^2 + 13x - 35$

15. If $101 = (99!)x$, then $x =$

(A) 1   (B) $\dfrac{101}{99}$ (C) 9,900   (D) 10,100

16. At a fruit market where no sales tax is charged, Abe paid $1.90 for 1 pear and 3 apples; Joan paid $1.60 for 1 pear, 1 apple, and 1 orange; and Latonya paid $1.70 for 2 apples and 1 orange. If each person paid the same amount per pear, the same amount per apple, and the same amount per orange, what amount did each person pay for each apple?

(A) $0.40   (B) $0.45   (C) $0.50   (D) $0.52

17. If the range of $f(x) = x^2 + 4$ is all real numbers from 13 to 29, what positive numbers lie in the domain of $f(x)$?

(A) $3 \leq x \leq 5$ (G) $5 \leq x \leq 21$ (C) $9 \leq x \leq 25$ (D) $13 \leq x \leq 29$

18. What is the 50th term of the arithmetic sequence 4, 10, 16, 22, $\cdots$?

(A) 202   (G) 206   (C) 294   (D) 298

19. If $f(x) = 3x + 2$, then $f(a+b) =$

(A) $3a + 3b + 2$ (B) $3a + 3b + 4$ (C) $3x + 2 + a + b$ (D) $3x + 4 + 3a + 3b$

20. If $i = \sqrt{-1}$, which of the following is equivalent to $\dfrac{2i}{1+i}$?

(A) -2 (B) $-1 + 2i$ (C) $1 - i$ (D) $1 + i$

21. Which of the following equations has $-i, i,$ and 0 as its only roots?

(A) $x^2 - 1 = 0$ (B) $x^3 + x = 0$ (C) $x^2 + x + 1 = 0$ (D) $x^3 - x = 0$

22. A sighting from sea level to the top of a lighthouse was 60°. The lighthouse is known to rise 180 feet above sea level. What is the distance (to the nearest foot) between the observer and the base of the lighthouse?

(A) 104   (B) 180   (C) 208   (D) 254

23. If $\sin x = \dfrac{1}{2}$ and x is between $\dfrac{\pi}{2}$ and $\dfrac{3\pi}{2}$, what is the value of $\dfrac{x}{2}$?

(A) $\dfrac{5\pi}{6}$ (B) $\dfrac{7\pi}{12}$ (C) $\dfrac{5\pi}{12}$ (D) $\dfrac{\pi}{12}$

24. Which of the following is another expression for $\dfrac{1+\csc\theta}{\sec\theta} - \cot\theta$?

(A) $\sin\theta + \tan\theta - \cot\theta$ (B) $\cos\theta + \tan\theta - \cot\theta$ (C) $\cos\theta$ (D) $\sin\theta$

25. If $\sin 2x = \sin x$, then which of the following could NOT be true?

(A) $-\dfrac{\pi}{3}$ (B) $\sin x = 0$ (C) $\cos x = 0$ (D) $\cos x = \dfrac{1}{2}$

26. $2i(4 - 6i) =$

(A) $8i + 12$ (B) $8i - 12$ (C) $12i - 8i$ (D) $8 - 12i$

27. 34. If for all x, $f(x) = x^2 - 2x + 3$ and $g(x) = x^2 - 3x + 4$, what is the value of $\dfrac{f(2)}{g(3)}$?

(A) $-\dfrac{1}{2}$ (B) $\dfrac{1}{2}$ (C) $\dfrac{2}{3}$ (D) $\dfrac{3}{4}$

28. If $f(x) = \dfrac{5x + 2}{3}$, what is the y-intercept of the graph of $f^{-1}(x)$?

(A) $-\dfrac{3}{2}$ (B) $-\dfrac{2}{5}$ (C) $\dfrac{3}{5}$ (D) $\dfrac{2}{3}$

29. At a convenience store, two candy bars and two bags of potato chips cost $4.00, and three candy bars and two bags of potato chips cost $4.75. What is the price of one bag of potato chips?

(A) $0.50   (B) $0.75   (C) $1.00   (D) $1.25

30. At a banquet of 36 people, every person had a choice among beef stroganoff, chicken divan, and linguini primavera. If 25% chose beef stroganoff and 17 people chose chicken divan, how many people chose linguini primavera?

(A) 7   (B) 8   (C) 9   (D) 10

31. What is the value of $4v(w2 - 3vw)$ given that $v = -1$ and $w = 4$?

(A) -84 (B) -92 (C) -104 (D) -112

32. Annette took 10 minutes to walk around a rectangular field. The length of the field is 4 times its width. How long would it take Annette to walk the width of the field?

(A) 1 minute   (B) 2 minutes   (C) 3 minutes   (D) 4 minutes

33. The graph is a parabola. For $(x,\ y)$ on the graph, the minimum value of y is attained at what value of x?

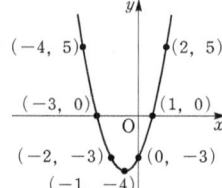

(A) -4
(B) -1
(C) -2
(D) 1

34. Two values of v satisfy the equation $|2v - 13| + 6 = 9$. What is the sum of these values?

(A) 4   (B) 7   (C) 9   (D) 13

35. The six squares in the figure have sides of lengths 2, 4, 6, 8, 10, and 12, respectively. What is the sum of the area of the shaded regions?

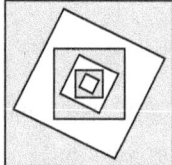

(A) 52
(B) 56
(C) 84
(D) 72

36. Which of the following functions has a range of $f(x) \geq 4$?

(A) $f(x) = |x + 4|$ (B) $f(x) = |x - 4|$ (C) $f(x) = |x + 4| - 4$
(D) $f(x) = |x - 4| + 4$

37. At a certain school, 200 juniors out of the total junior class of 250 students attended the last football game of the season. If 220 freshmen also attended that game, and if the fraction of freshmen who attended was equal to the fraction of juniors who attended, what was the total number of students in the freshmen class?

(A) 120
(B) 275
(C) 225
(D) 250

38. If $\dfrac{a^2+2ab+b^2}{a^2-b^2}=2(a+b)$, what is the value of $a-b$?

(A) 1   (B) $-\dfrac{1}{2}$ (C) 2   (D) $\dfrac{1}{2}$

39. Which of the following is a possible solution for x in terms of k for the equation $x=\dfrac{2k}{x+2}$?

(A) $\sqrt{2k}$ (B) $\sqrt{-2k}$ (C) $\sqrt{1+2k}+1$ (D) $\sqrt{1+2k}-1$

40. If $49^{3y}=\sqrt{7^{y+1}}$, then $y=$

(A) $\dfrac{1}{3}$ (B) $\dfrac{1}{5}$ (C) $\dfrac{1}{7}$ (D) $\dfrac{1}{11}$

41. If $\dfrac{f}{g}=\dfrac{1}{4}$ and $\dfrac{g}{h}=\dfrac{2}{5}$, what is the ratio of $f:h$?

(A) 1:6   (B) 1:8   (C) 1:10   (D) 10:1

42. A garden consists of a continuous chain of flower beds in the shape of hexagons, the beginning of which is shown in the figure. There are 15 flower beds in the chain, and each one, except the first and last, shares two of its sides with adjacent flower beds. If the length of each side of each bed is 1 meter, what is the perimeter of the garden?

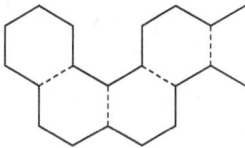

(A) 62 meters (B) 58 meters
(C) 59 meters (D) 60 meters

43. Which of the following is equivalent to $\dfrac{\tan x \cdot \csc x}{\sin x \cdot \sec x}$?

(A) $\sin x$ (B) $\cos x$ (C) $\cot x$ (D) $\csc x$

44. Martha picked out a pair of shoes she wanted and brought them to the front of the store to pay. The cashier told her that the shoes were on sale for 30% off the original price. She rang up the sale price plus 5% sales tax, so Martha ended up paying $58.80 for the shoes. What was the original price for the shoes?

(A) $70   (B) $75   (C) $80   (D) $85

45. In a certain election, 65 percent of those who voted were females. If 7,000 males voted, what was the total number of people who voted in the election?

(A) 12,000
(B) 14,000
(C) 18,600
(D) 20,000

> Use the following information to answer Questions 46: Eldridge opened a savings account with an initial balance of $1,000. After that, every month for 8 months, she made one deposit to the account, always for the same amount. During that time, she made just one withdrawal.

46. If each of Eldridge's 8 deposits was in the amount of $200 and her withdrawal was $350, how much money did her account have at the end of 8 months?

(A) $850   (B) $1,550   (C) $2,250   (D) $2,350

47. In the figure, O is the center of the circle and $\overline{AD}$, $\overline{BE}$, and $\overline{CF}$ are diameters. If the length of arc $\overset{\frown}{AB}$ is 18, what is the length of arc $\overset{\frown}{DEF}$?

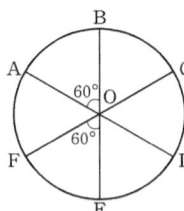

(A) 24
(B) 30
(C) 36
(D) 48

48. What is the value of x if $\dfrac{2x+3y-19}{y+5}=3$?

(A) 17   (B) −7   (C) 29   (D) −9

49. A scientist performs an experiment in which she measures four values two times each, with the following results:

	w	x	y	z
First Measurement	0.2	6	0.5	10
Second Measurement	0.6	3	1	30

Which of the following conclusions does the experiment provide evidence for?

(A) w and y are directly proportional.   (B) w and z are inversely proportional.

(C) x and y are inversely proportional.   (D) x and z are inversely proportional.

50. Which of the following could be the equation of the graph?
   (A) $y=(x-2)^2+2$
   (B) $y=(x+2)^2-2$
   (C) $y=(x-2)^2-2$
   (D) $y=x^2+2$

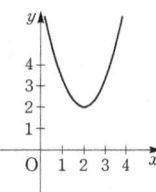

51.

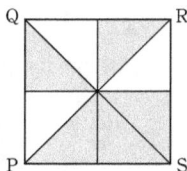

In the figure above, PQRS is a square
What percent of the square is shaded?

(A) $37\frac{1}{2}\%$ (B) 50%   (C) 60%   (D) $62\frac{1}{2}\%$

52.

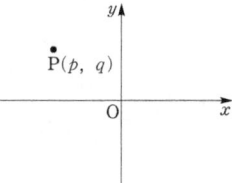

In the figure above, point Q is the same distance from O as point P is from O. Which of the following could be the coordinates of point Q?

(A) $(-p,\ q)$ (B) $(q,\ p)$ (C) $(-q,\ -p)$ (D) $(q,\ -p)$

53. If $g(x)$ is a transformation that moves $f(x)$ three units to the right and then reflects it across the x-axis, then $g(x) =$

(A) $f(-x) + 3$ (B) $f(-x) - 3$ (C) $-f(x + 3)$ (D) $-f(x - 3)$

54. Which of the following is the domain of the function $f(x) = \dfrac{3 - x}{\sqrt{x^2 - 9}}$?

(A) $-3 \le x \le 3$ (B) $-3 < x < 3$ (C) $x < -3$ or $x > 3$ (D) $x \le -3$ or $x \ge 3$

55. A two-digit number from 10 to 99, inclusive, is chosen at random. What is the probability that this number is divisible by 5?

(A) $\dfrac{1}{5}$ (B) $\dfrac{2}{9}$ (C) $\dfrac{19}{90}$ (D) $\dfrac{18}{91}$

56. If A can do a job in 8 days and B can do the same job in 12 days, how long would it take the two men working together?

(A) 3   (B) 4.8   (C) 10   (D) 5.8

57. An isosceles triangle contains three angles that measure $40°$, $x°$, and $y°$. Which of the following CANNOT be true?

(A) $x = y$ (B) $x = 50$ (C) $x - y = 60$ (D) $x = 70$

58.

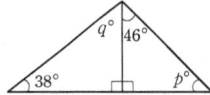

In the figure above, what is the value of $p + q$?

(A) 94   (B) 96   (C) 106   (D) 114

59. In the complex numbers, where $i^2 = -1$, what is the value of $5 + 6i$ multiplied by $3 - 2i$?

(A) 27   (B) 27i (C) $27 + 8i$ (J) $15 + 8i$

60. What is the equation for a line that intersects the origin and is perpendicular to $2x - 4y = 13$?

(A) $y = -2x$ (B) $y = 2x$ (C) $y = \dfrac{1}{2}x$ (D) $y = \dfrac{1}{2}x - \dfrac{13}{4}$

61. This semester, Gerry scored an average of 93 on his five history exams. He got the same score on his first two exams, and then he got a 94, an 85, and a 90 on the remaining exams. What score did he receive on his first two exams?

(A) 95   (B) 96   (C) 97   (D) 98

62. On an xy-graph, what is the length of a line segment drawn from $(-3, 7)$ to $(6, -5)$?

(A) 15   (B) 16   (C) 17   (D) 18

63. Which of the following is NOT a factor of $4x^2y^4 - 12x^3y^2 - 8xy^2$?

(A) $2x^2$ (G) $2xy^2$ (C) $-2y$ (D) $4y^2$

64. Of the 126 students who applied for a full scholarship at Oxbow College, 9 received one. What is the ratio of students who received a scholarship to those who didn't?

(A) 1 to 10   (B) 1 to 11   (C) 1 to 12   (D) 1 to 13

65. The length of the sides of an isosceles triangle are 24, k, and k. If k is an integer, what is the smallest possible perimeter of the triangle?

(A) 48
(B) 49
(C) 52
(D) 50

66. If $2x - y = 32$ and $5x + 3y = 14$, then $xy =$

(A) 35 (B) 75 (C) 100 (D) −120

67. What is the value of a in terms b if $\dfrac{a}{b} + \dfrac{a+2}{3b} = \dfrac{1}{4}$?

(A) $\dfrac{b+4}{2}$ (B) $\dfrac{3b-2}{4}$ (C) $\dfrac{3b+8}{16}$ (D) $\dfrac{3b-8}{4}$

68. If p percent of 250 is 75, what is 75% of p?

(A) 22.5 (B) 25 (C) 75 (D) 225

69. What is the slope of the line $9x - 3y = 10$?

(A) 3 (B) −3 (C) 9 (D) $\dfrac{1}{3}$

70. Alfred and Rani both picked different two digit numbers. If you multiply Alfred's number by 5 and double Rani's number, the sum is 300. If you double Alfred's number and multiply Rani's number by 3, the sum of the two numbers is 252. What is the sum of their two numbers?

(A) 96 (B) 112 (C) 128 (D) 144

71.

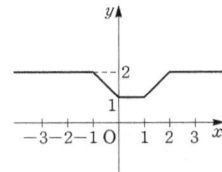

The graph of $y=f(x)$ is shown above. If $f(x)=1$, then p could be equal to

(A) -1.5 (B) -0.5 (C) 1.5 (D) 0.5

72. The formula for the volume of a sphere is $V=\dfrac{4}{3}\pi r^3$, and the formula for the surface area of a sphere is $A=4\pi r^2$. If a sphere has a surface area of 36π, what is its volume?

(A) 27π (B) 36π (C) 54π (D) 72π

73. What is the sum of the two values of x that satisfy the equation $4x^2-3x-1=0$?

(A) 0.75 (B) 1.25 (C) 2.5 (D) $-.75$

74. What is a possible value of x if $\sqrt{2x+3}-1=x$?

(A) $\sqrt{2}$ (B) $\sqrt{3}$ (C) $\sqrt{5}$ (D) $\sqrt{7}$

75. If $|6-4n|>1$, which of the following must be true?

(A) $\dfrac{5}{4}<n<\dfrac{7}{4}$ (B) $-\dfrac{5}{4}<n<\dfrac{7}{4}$ (C) $n<\dfrac{5}{4}$ or $n>\dfrac{7}{4}$ (D) $n>\dfrac{5}{4}$ or $n\leftarrow\dfrac{7}{4}$

76. If a and b are constants, what is the maximum number of point of intersection between the graph of $y=|ax^2-8x+b|$ and a circle?

(A) 2 (B) 4 (C) 6 (D) 8

77. Which of the following equations expresses y in terms of x for each of the four pairs of values shown in the table ?

x	y
1	6.5
2	14.0
3	21.5
4	29.0

(A) $y=6.5x+7.5$
(B) $y=7.5x-1$
(C) $y=7.5x+6.5$
(D) $y=5.5x+1$

78. A clown at an amusement park makes animal shapes from twisted balloons. She sells each animal based on the number of balloons it requires, according to the following chart: **Number of balloons:**

	1	2	3	4	5
Price	$4.00	$4.50	$5.00	$5.50	$6.00

Which of the following functions equals the dollar cost of a balloon animal that contains x balloons?

(A) $f(x) = x$ (B) $f(x) = x + 4$ (C) $f(x) = 0.5x + 4$ (D) $f(x) = 0.5x + 3.5$

79. When $x = 3$ and $y = 5$, by how much does the value of $3x^2 - 2y$ exceed the value of $2x^2 - 3y$?
(

A) 4   (B) 14   (C) 16   (D) 20

80. Sales for a business were 3 million dollars more the second year than the first, and sales for the third year were double the sales for the second year. If sales for the third year were 38 million dollars, what were sales, in millions of dollars, for the first year?

(A) 16   (B) 17.5   (C) 20.5   (D) 22

81. Abandoned mines frequently fill with water. Before an abandoned mine can be reopened, the water must be pumped out. The size of pump required depends on the depth of the mine. If pumping out a mine that is D feet deep requires a pump that pumps a minimum of $\frac{D^2}{25}+4D-250$ gallons per minute, pumping out a mine that is 150 feet deep would require a pump that pumps a minimum of how many gallons per minute?

(A) 362   (B) 500   (C) 800   (D) 1,250

82. A typical high school student consumes 67.5 pounds of sugar per year. As part of a new nutrition plan, each member of a track team plans to lower the sugar he or she consumes by at least 20% for the coming year. Assuming each track member had consumed sugar at the level of a typical high school student and will adhere to this plan for the coming year, what is the maximum number of pounds of sugar to be consumed by each track team member in the coming year?

(A) 14   (B) 44   (C) 48   (D) 54

83. For which of the following ordered pairs (p, q) is $p+q>5$ and $p-q<-6$?

(A) $(1, 8)$ (B) $(6, 5)$ (C) $(5, 6)$ (D) $\left(\frac{1}{2}, \frac{9}{2}\right)$

84. If a discount of 25% off the retail price of a desk saves Mark $45, how much did he pay for the desk?

(A) $135   (B) $160   (C) $180   (D) $210

85. A customer pays $1,100 in state taxes on a newly purchased car. What is the value of the car if state taxes are 8.9% of the value?

(A) $9.765.45   (B) $10,876.90   (C) $12,359.55   (D) $14,345.48

86. How many years does Steven need to invest his $3,000 at 7% to earn $630 in simple interest?

(A) 1 year   (B) 2 years   (C) 3 years   (D) 4 years

87. Sabrina's boss states that she will increase Sabrina's salary from $12,000 to $14,000 per year if she enrolls in business courses at a local community college. What percent increase in salary will result from Sabrina taking the business courses?

(A) 15%   (B) 16.7%   (C) 17.2%   (D) 85%

88. Jim works for $15.50 per hour for a health care facility. He is supposed to get a 75 cent per hour raise at one year of service. What will his percent increase in hourly pay be?

(A) 2.7%   (B) 3.3%   (C) 133%   (D) 4.8%

89. If 45 is 120% of a number, what is 80% of the same number?

(A) 30   (B) 32   (C) 36   (D) 38

90. How long will Lucy have to wait before her $2,500 invested at 6% earns $600 in simple interest?

(A) 2 years   (B) 3 years   (C) 4 years   (D) 5 years

91. What is 35% of a number if 12 is 15% of a number?

(A) 5   (B) 12   (C) 28   (D) 33

92. A computer is on sale for $1600, which is a 20% discount off the regular price. What is the regular price?

(A) $1800   (B) $1900   (C) $2000   (D) $2100

93. A car dealer sells a SUV for $39,000, which represents a 25% markup over the dealer's cost. What was the cost of the SUV to the dealer?

(A) $29,250   (B) $31,200   (C) $32,500   (D) $33,800

94. After having to pay increased income taxes this year, Edmond has to sell his BMW. Edmond bought the car for $49,000, but he sold it for a 20% loss. What did Edmond sell the car for?

(A) $28,900   (C) $35,600   (C) $37,300   (D) $39,200

95. Lauren had $80 in her savings account. When she received her paycheck, she made a deposit which brought the balance up to $120. By what percentage did the total amount in her account increase as a result of this deposit?

(A) 50%   (B) 40%   (C) 35%   (D) 80%

96. What are the solutions to the quadratic equation $x^2 - 15x + 36 = 0$

(A) 3, 6   (B) 4, 9   (C) 6, 6   (E) −4, −9

97. If for all x and y, $2x - b - y = 0$, then $b = ?$

(A) $2x - y$ (B) $y - 2x$ (C) $2y + x$ (D) $2y - x$

98. For all $x \neq 4$, $\dfrac{x^2 - 8x + 16}{x - 4}$?

(A) $x + 4$ (B) $x - 4$ (C) $x + 8$ (D) $x - 8$

99. What is the value of the expression $4x\sqrt{3x} + 5x\sqrt{8x}$ when $x = 2$

(A) $8\sqrt{6} + 40$ (B) $6\sqrt{5} + 7\sqrt{10}$ (C) $48\sqrt{6}$ (D) $166\sqrt{6}$

100. If $x \geq -1$, $4 + \sqrt{5x + 5} = 9$, then $x = ?$

(A) 0   (B) 25   (C) 4   (D) 9

101. For all $x \geq 0$ and $y \geq 0$, $\sqrt{250x^9y^4} = ?$

(A) $3x\sqrt{28y^7}$ (B) $5x^4y^2\sqrt{10x}$ (C) $10x^2y^2\sqrt{5x^2y}$ (D) $5x^3y^2\sqrt{3}$

102. Simplify $\dfrac{\sqrt{32}}{2} + \dfrac{2\sqrt{3}}{3} = ?$

(A) $4 + 2\sqrt{3}$ (B) $4 + \dfrac{2\sqrt{3}}{3}$ (C) $\dfrac{6\sqrt{2} + 2\sqrt{3}}{2}$ (D) $\dfrac{2\sqrt{35}}{5}$

103. If $\dfrac{a}{b} = \dfrac{7}{4}$ and $\dfrac{b}{c} = \dfrac{4}{5}$, then $\dfrac{a}{c} =$

(A) $\dfrac{16}{35}$ (B) $\dfrac{5}{7}$ (C) $\dfrac{7}{5}$ (D) $\dfrac{35}{16}$

104. In the figure, the slope of line l is $\frac{3}{4}$. What is the value of p?

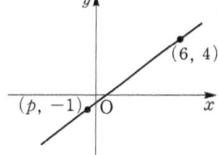

(A) $-\frac{2}{3}$ (B) $-\frac{5}{2}$ (C) $\frac{2}{3}$ (D) 2

105. For all x, $3(2x+5)-4(x-2)=3(2x+2)+1$

(A) 9   (B) 3   (C) 4   (D) -5

106. A cellphone plan costs \$2 per month plus 10 cents per minute of talk time used. If the total talk time used in a month is m minutes, what is the total cost in dollars?

(A) $10+2m$ (B) $2+10m$ (C) $0.1+2m$ (D) $2+0.1m$

107. Suppose $x+y=17$. What is $(x+y)^2-3(x+y)+12$?

(A) 50   (B) 100   (C) 150   (D) 200

108. A jar with capacity 3 gallons is used to fill a tank with water. In each trip, the jar is partly or fully filled with water from a faucet and the jar is then taken to the tank where the water is emptied into the tank. If the tank's capacity is 91 gallons, what is the minimum number of trips needed to fill the tank?

(A) 30   (B) 31   (C) 88   (D) 94

109. Cold drink cans be bought either individually for \$ 1.30 a can or in packs of 15 for \$ 18 a pack. What is the minimum amount of money needed to purchase a total of 40 cans? (We are allowed to buy some packs and some individual cans.

(A) \$ 36   (B) \$ 49   (C) \$ 52   (D) \$ 54   (E) \$ 60

110 . If $a^2 + 14a = 51$ and $a > 0$, what is the value of $a + 7$?

111. The function f is defined by $f(x) = 2x^3 + 3x^2 + cx + 8$ where c is a constant. In the xy-plane, the graph of f intersects the x-axis at the three points $(-4, 0), (\frac{1}{2}, 0)$ and $(p, 0)$. What is the value of c ?

(A) -18 (B) -2 (C) 2   (D) 10

112.

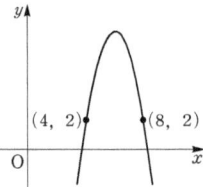

The figure above shows the graph of a quadratic function in the xy-plane. Of all the points (x, y) on the graph, for what value of x is the value of y greatest?

(A) 2   (B) 6   (C) 5   (D) 7

113. What is one possible solution to the equation $\dfrac{24}{x+1} - \dfrac{12}{x-1} = 1$?

114. If $f(x) = ax^2 + bx + c$ is concave downward, which of the following must be true?

(A) $a < 0$ (B) $a > 0$ (C) $b^2 - 4ac > 0$ (D) $b^2 - 4ac < 0$

115. Anise needs to complete a printing job using both of the printers in her office. One of the printers is twice as fast as the other, and together the printers can complete the job in 5 hours. The equation above represents the situation described. Which of the following describes what the expression $\frac{1}{x}$ represents in this equation?

$$\frac{1}{x}+\frac{2}{x}=\frac{1}{5}$$

(A) The time, in hours, that it takes the slower printer to complete the printing job alone

(B) The portion of the job that the slower printer would complete in one hour

(C) The portion of the job that the faster printer would complete in two hours

(D) The time, in hours, that it takes the slower printer to complete $\frac{1}{5}$ of the printing job

116. If $f(x)=\dfrac{x+1}{x^2-x-30}$, the vertical asymtote(s) of $f(x)$ is(are)?

(A) $x=10,-3$ (B) $x=6$ (C) $x=-5,6$ (D) $x=-6,5$

117. If (x,y) is a solution to the system of equations below, what is the value of x^2?

$$x^2+y^2=153$$
$$y=-4x$$

(A) −51   (B) 3   (C) 9   (D) 144

118. The population of a region in the United States was 10.7 million. If the population were to increase 0.43% annually for 17 years, by what percentage would the population have increased after such a period?

(A) 4.97% (B) 5.77% (C) 6.32% (D) 7.57%

119. If $3^{(x-\frac{1}{5})^2} = 1$, then $x =$

(A) $\dfrac{1}{5}$ (B) $\dfrac{1}{3}$ (C) 3 (D) 5

120. Which of the following is always true for all real numbers?

I. $x^2 - 2x - 3 > 0$ II. $x^2 - 4x + 4 > 0$ III. $x^2 - x - 5 > 0$

(A) I (B) II (C) III (D) I, II

121. If x is an acute angle and $\cos\left(\dfrac{\pi}{2} - 2x\right) = \dfrac{1}{2}$, then $x =$

(A) $\dfrac{\pi}{12}$ (B) $\dfrac{5}{12}\pi$ (C) $\dfrac{\pi}{12}, \dfrac{5}{12}\pi$ (D) $\dfrac{\pi}{12}, \dfrac{7}{12}\pi$

122. If $f(x) = a(x-2)^2 + 15$ and $f(1) = 2$, then $a =$

(A) -13 (B) -10 (C) 10 (D) 13

123. Which of the following is the solution set of $|x^3 - 3| < 7$?

(A) $-2 < x < 3.16$ (B) $-1.59 < x < 2.15$ (C) $x > 2.15$ or $x < -1.59$
(D) $x > 3.16$ or $x > 2$

124. If x and $f(x)$ are given in the table below and $f(f(x)) = 10$, then $x =$

x	3	5	7	8	10
$f(x)$	10	2	3	12	1

(A) 3   (B) 5   (C) 7   (D) 8

125. If $a^2 + b^2 = 15$ and $ab = 3.75$, then $(a+b)^2 =$

(A) 11.5   (B) 13.5   (C) 18.5   (D) 22.5

126. The height, in feet, of a stone thrown with an upward speed of 40 ft/s is given by the formula $h = 40t - 16t^2$, where t is the time, in seconds, since the stone was thrown. How long does it takes the stone to hit the ground?

127. Peter made a rectangular pen for his dog using a side of the for one side and 24 m of fencing for the remaining three sides. If the area enclosed was 72 m^2, find the dimensions of the pen.

128. If $a_1 = 1, a_2 = 3$, and $a_n = \dfrac{2a_{n-1} + 3}{a_{n-2}}$, then $a_5 =$

(A) 9   (B) 7   (C) $\dfrac{13}{6}$ (D) $\dfrac{17}{9}$

129. If $(x-1)(x-3)(x+3)(x+5) < 0$, the solution set is included by

(A) $-6 < x < -2$ (B) $-4 < x < -2$ (C) $-3 < x < 1$ (D) $0 < x \leq 1$

130. There is a cylinder. If its radius is decreased by 50% and the height is increased by 60% compared to its initial state, the volume of the cylinder _____

(A) increased by 40%   (B) decreased by 40%   (C) increased by 60%

(D) decreased by 60%

131. If $f(x, y, z) = (x, z, y)$ and $g(x, y, z) = (-z, -x, 0)$, then $f(g(3, 2, 1)) =$

(A) $(-3, -1, 0)$ (B) $(-3, 0, -1)$ (C) $(-1, -3, 0)$ (D) $(-1, 0, -3)$

132. Which of the following quadratic equations has a and b as its roots if $a+3$ and $b+3$ are the roots of $x^2 - 5x + 3 = 0$?

(A) $x^2 - 3x + 5 = 0$ (B) $x^2 + x - 5 = 0$ (C) $x^2 + 5x - 3 = 0$ (D) $x^2 + x - 3 = 0$

133. If $f(x) = 3x + 1$ and $(f \circ g)(x) = x$, then $g(x) =$

(A) $\dfrac{x+1}{3}$ (B) $\dfrac{x-1}{3}$ (C) $\dfrac{x}{3} - 1$ (D) $3x - 1$

134. If $|x| + |y| \leq 1$, which of the following can be (x, y)?

(A) $\left(-\dfrac{1}{3}, \dfrac{1}{2}\right)$ (B) $\left(-\dfrac{2}{3}, \dfrac{1}{2}\right)$ (C) $\left(\dfrac{3}{5}, -\dfrac{1}{2}\right)$ (D) $\left(-\dfrac{1}{2}, -\dfrac{2}{3}\right)$

135. If $y = 2x - 5$, which of the following has the same x-intercept?

(A) $y = \dfrac{1}{2}x + 10$ (B) $y = \dfrac{1}{3}x - \dfrac{10}{3}$ (C) $y = -\dfrac{2}{3}x + 5$ (D) $y = -\dfrac{2}{3}x + \dfrac{5}{3}$

136. If $y = ax + b$ and $y = cx + d$ are perpendicular, which of the following is true?

(A) $ac = -1$ (B) $a = c, b \neq d$ (C) $a = c, b = d$ (D) $a \neq d$

137. Which of the following has the same root when you replace x with $-x$?

(A) $2x^2 + 2x - 1 = 0$ (B) $2x^3 - 3x = 0$ (C) $x^4 + x^2 - 2x = 0$

(D) $x^4 - 3x^2 + x + 1 = 0$

138. A bus charge $0.30 for the first 1 mile and $0.15 for additional 1 mile. If a customer rides in this bus for x miles, which of the following expressions describes that cost, $f(x)$, of this ride in dollars?

(A) $f(x) = 0.30 + 0.15x$ (B) $f(x) = 0.30 + 0.15(1 - x)$ (C) $f(x) = 0.30 + 0.15(x - 1)$

(D) $f(x) = 0.30 + 0.15x - 1$

139. A student begins to walk on a road directly towards a tree that is 20 feet tall. How far does the student travel during the time that the angle of elevation from the student to the top of tree changes from $18°$ to $31°$? (Disregard the student's height)

(A) 28.268 feet   (B) 30.012 feet   (C) 32.315 feet   (D) 35.703

140. Write without radicals in the denominator: $\dfrac{\sqrt{3}+1}{\sqrt{3}-1}$

(A) $2 + 2\sqrt{3}$ (B) $2 - 2\sqrt{3}$ (C) $2 - \sqrt{3}$ (D) $2 + \sqrt{3}$

141. A piece of wire 47 cm long is cut so that the two pieces can be formed into a square and an equilateral triangle. The sum of the lengths of a side of the square and a side of the triangle is 15 cm. How long is the side of the square?

(A) 1 cm   (B) 2 cm   (C) 8 cm   (D) 9.75 cm

143. If $1000 is invested at 6% interest compounded annually, how many years will it take (to the nearest year) for the investment to double?

(A) 12   (B) 14   (C) 16   (D) 17

144. Solve for x: $\dfrac{5}{x+1} \geq 1$

(A) $x \leq -4$ or $x > 1$ (B) $-1 < x \leq 4$ (C) $x \leq -1$ or $x > 4$
(D) $-4 < x \leq 1$

145. An equilateral triangle is inscribed in a circle with radius 2 meters. What is the area of this triangle?

(A) $\sqrt{3}$ (B) $2\sqrt{3}$ (C) $3\sqrt{3}$ (D) $\pi\sqrt{3}$

146. Two companies produce equivalent CD players at the same production cost. They sell the same number of units in each 6-month period at current selling price of $100. The first company plans to reduce its selling price by 5% at the end of each 6-month period, and the second plans to reduce its price by 11% after each year. What will be the positive difference in their prices (to the nearest cent) five years from now?

(A) $22.54   (B) $21.54   (C) $5.03   (D) $4.03

147. Solve for x: $-5 < \dfrac{1}{2}x - 3 < -1$

148. An opinion poll asked which of two candidates, A or B, would make a good mayor. Of respondents, 70% chose A and 60% chose B. Each person polled chose at least one, and 900 of them chose both candidates. How many people were polled?

(A) 1000   (B) 2000   (C) 3000   (D) 4000

149. If $f(x)=2\cos x$ and $g(x)=x^4-2$, then the value of $g(f(\frac{\pi}{6}))=$

(A) 7   (B) 9   (C) 12   (D) 18

150. Find the sum of the roots of $|x-7|^2+2|x-7|=24$.

(A) 4   (B) 12   (C) 14   (D) 16

151. If the expression $\dfrac{4x^2}{2x-1}$ is written in the equivalent form $\dfrac{1}{2x-1}+A$, what is A in terms of x?

(A) $2x+1$ (B) $2x-1$ (C) $4x^2$ (D) $4x^2-1$

152. Aaron is staying at a hotel that charges \$99.95 per night plus tax for a room. A tax of 8% is applied to the room rate, and an additional one-time untaxed fee of \$5.00 is charged by the hotel. Which of the following represents Aaron's total charge, in dollars, for staying x nights?

(A) $(99.95+0.08x)+5$ (B) $1.08(99.95x)+5$ (C) $1.08(99.95x+5)$
(D) $1.08(99.95+5)x$

153. Find all solutions of the radical expression $\sqrt{2x+3} + \sqrt{x+1} = 1$

A) $\{-1, 3\}$ 　　　(B) $\{1, -3\}$ 　　　(C) $\{-1\}$ 　　　(D) $\{3\}$

154. The gas mileage for Peter's car is 21 miles per gallon when the car travels at an average speed of 50 miles per hour. The car's gas tank has 17 gallons of gas at the beginning of a trip. If Peter's car travels at an average speed of 50 miles per hour, which of the following functions f models the number of gallons of gas remaining in the tank t hours after the trip begins?

(A) $f(t) = 17 - \dfrac{21}{50t}$ 　　　(B) $f(t) = 17 - \dfrac{50t}{21}$ 　　　(C) $f(t) = \dfrac{17 - 21t}{50}$

(D) $f(t) = \dfrac{17 - 50t}{21}$

155. If $-\dfrac{9}{5} < -3t+1 < -\dfrac{7}{4}$, what is one possible value of $9t-3$?

156. In the equation below, what is the value of k?
$$\frac{5(k+2)-7}{6} = \frac{13-(4-k)}{9}$$

(A) $\dfrac{9}{17}$ 　　　(B) $\dfrac{9}{13}$ 　　　(C) $\dfrac{33}{17}$ 　　　(D) $\dfrac{33}{13}$

157. One store sold red pens at four for a dollar and yellow pens at three for a dollar. A second store sold red pens at four for a dollar and yellow pens at six for a dollar. You bought m red pens and n yellow pens from each store, spending a total of ten dollars. How many pens in all did you buy?

(A) 10 　　　(B) 20 　　　(C) 30 　　　(D) 40

158. The solution of the system $3x+2y=-1$ is:
$$6x+4y=-2$$

(A) $y=x$ (B) No solution   (C) $(1,-2)$ (D) None of these

159. Water is flowing into your basement at a rate of 50 liters per hour. You can bail the water out at a rate of 60 liters per hour. If the water has been flowing in for two hours before you begin bailing, how many hours would it take you to bail out all the water from your basement?

(A) 10   (B) 20   (C) 30   (D) 40

160. Find an equation for the line that passes through the point (1, 2) and is perpendicular to the line $x+y=2$.

(A) $y=1-x$ B) $y=3-x$ (C) $y=1+x$ (D) $y=2+x$

161. Based on the system of equations below, what is the value of the product xy?
$$4x-y=3y+7$$
$$x+8y=4$$

(A) $-\dfrac{3}{2}$ (B) $\dfrac{1}{4}$ (C) $\dfrac{1}{2}$ (D) $\dfrac{11}{9}$

162. If $\dfrac{1}{2}x+\dfrac{1}{3}y=4$, what is the value of $3x+2y$?

163. The toll rates for crossing a bridge are \$6.50 for a car and \$10 for a truck. During a two-hour period, a total of 187 cars and trucks crossed the bridge, and the total collected in tolls was \$1,338. Solving which of the following systems of equations yields the number of cars, x, and the number of trucks, y, that crossed the bridge during the two hours?

(A) $x+y=1338$
$6.5x+10y=187$

(B) $x+y=187$
$6.5x+10y=\dfrac{1338}{2}$

(C) $x+y=187$
$6.5x+10y=1338$

(D) $x+y=187$
$6.5x+10y=1338\times2$

164. Sarah, Greg, and Heidi live in three houses along a straight road and Greg lives between Sarah and Heidi. Sarah leaves her house and walks toward Greg's house. When Sarah is $\dfrac{2}{3}$ of the way to Greg's house, she is $\dfrac{1}{2}$ of the way to Heidi's house.

Suppose Heidi leaves her house and walks toward Sarah's house. When Heidi is $\dfrac{2}{3}$ of the way to Greg's house, what fraction of the way to Sarah's house is she?

(A) $\dfrac{1}{2}$ (B) $\dfrac{1}{3}$ (C) $\dfrac{1}{4}$ (D) $\dfrac{1}{6}$

165. Find the number of digits of $2^{2005} \cdot 5^{2003} - 1$.

(A) 2002   (B) 2003   (C) 2004   (D) 2005

166. A theater has 36 rows and each row has 32 seats. The seats are numbered from 1 to 1152, beginning with the first row. In which row is the seat with number 857?

(A) 24   (B) 25   (C) 26   (D) 27

167. Which of the following equations represent the line tangent to the circle $x^2 + y^2 = 25$ at the point (3,4)?

(A) $3x + 4y - 25 = 0$ (B) $3x - 4y - 25 = 0$ (C) $4x + 3y + 3 = 0$

(D) $4x - 3y = 0$

168. An arch, shaped like a parabola, is 30 m wide at the ground base and 15 m high. How wide is the arch 10 m from the ground? Round your answer to the nearest hundredth of a meter.

(A)19.34m   (B)18.64m   (C)18.12m   (D)17.32m

169. A small tree 5 feet from a lamp post casts a shadow 4 feet long. If the lamp post were 2 feet taller the shadow would only be 3 feet long. How tall is the tree?

(A) $\dfrac{9}{2}$ ft.   (B) $\dfrac{16}{3}$ ft.   (C) $\dfrac{14}{3}$ ft.   (D) $\dfrac{24}{5}$ ft.

170. The area of a rectangle is $2x^4 + 5x^3 - 20x^2 - 16x + 35$ and its width is $x^2 + 3x - 5$. The length of the rectangle is

(A) $2x^4 + 5x^3 + 5x^2 + 2x - 7$ (B) $2x^3 + 5x^2 + 2x - 16x - 7$

(C) $2x^2 + 5x - 7$ (D) $2x^2 - x - 7$

171. How many distinct solutions of the form (x, y), where both x and y are integers, does the following equation have?

$$x^2 \cdot (|y| + 3) = 12$$

(A) 2   (B) 4   (C) 6   (D) No solution

172. The product of the values of m and n that require the graph of $y = x^2 + mx + n$ to pass through the points $(1,12)$ and $(3, 28)$ is

(A) 20   (B) 36   (C) -40 (D) 28

173. Let $f(x)$ be a function such that $f(x) = 2f(x-1) + 4$ and $f(0) = 16$. What is the value of $f(-2) + f(2)$?

(A) 0   (B) 16   (C) 32   (D) 77

174. In the system of linear equations below, a is a constant. If the system has no solution, what is the value of a?

$$\frac{1}{2}x - \frac{1}{4}y = 5$$
$$ax - 3y = 20$$

(A) $\dfrac{1}{2}$ (B) 2   (C) 6   (D) 12

175. Solve completely for x: $3^{x^2} = 9^8$

(A) -2 , 2   (B) -4 , 4   (C) 8   (D) -16 , 16

176. The number of distinct real values x which satisfy the equation $(x^2 - 6x + 9)^{(x^2 - 8x + 12)} = 1$ is:

(A) 0   (B) 1   (C) 2   (D) 3

177. Solve for t: $|\frac{2}{3}t - 2| = |\frac{1}{3}t + 3|$

(A) $1, -15$ (B) $-1, 15$ (C) $5, -\dfrac{1}{3}$ (D) 15

178. How many real zeros does the function $f(x) = x^4 + x^3 - 2x^2 + 4x - 24$ have?

(A) 0   (B) 1   (C) 2   (D) 3

179. Let $f(x) = x^2 - 4x$ and $g(x) = 3x - 1$. Define the operation $f \triangle g$ by $(f \triangle g)(x) = f(g(x)) - g(f(x))$. Then $(f \triangle g)(1) =$

(A) -6 (B) -5 (C) 0   (D) 6

180. Find the area of the circle that passes through the points (0,0), (0,4), and (4,0).

(A) 3π (B) $2\sqrt{2}\,\pi$ (C) 8π (D) 16π

181. The sum of the coordinates of the point of intersection of the lines

$$\frac{x}{5} + 2y = \frac{16}{5} \quad \text{is:}$$
$$\frac{3x}{5} + \frac{y}{2} = -\frac{7}{5}$$

(A) -4 (B) -2 (C) 2   (D) 4

182. The regular price for a T-shirt is $25 and the regular price for a pair of jeans is $75. If the T-shirt is sold at a 30% discount and the jeans are sold at a 10% discount, then the total discount is

(A) $15   (B) $20   (C) $30   (D) $36   (E) $40

183. Let m be an even integer. How many possible values of m satisfy $\sqrt{m+7} \leq 3$?

(A) One   (B) Two   (C) Three   (D) Four   (E) Five

184. Going into the final exam, which will count as two tests, Courtney has test scores of 80, 81, 73, 65 and 91. What score does Courtney need on the final in order to have an average score of 80?

(A) 83   (B) 84   (C) 85   (D) 86

185. If A, B, C are constants such that for all values of x,
$x^2 - x - 2 = (Ax + B)(x - 2) + C(x^2 + 3)$, what is the value of A?
(A) 1   (B) 2   (C) 3   (D) 4 E.

186. Let $\boxed{x}$ be defined by $\boxed{x} = \dfrac{x+3}{x-1}$ for any x such that $x \neq 1$. Which of the following is equivalent to $\boxed{x} - 1$?

(A) $\dfrac{x+2}{x-1}$ (B) $\dfrac{4}{x-1}$ (C) $\dfrac{2x+4}{x-1}$ (D) $\dfrac{2}{x-1}$ (E) $\dfrac{x+2}{x-2}$

187. The manager of a store that specializes in selling tea decides to experiment with a new blend. She will mix some Earl Grey tea that sells for $5 per pound with some Orange Pekoe tea that sells for $3 per pound to get 100 pounds of the new blend. The selling price of the new blend is to be $4.50 per pound and there is to be no difference in revenue from selling the new blend versus selling the other types. How many pounds of the Earl Grey tea are required?

(A) 70   (B) 75   (C) 80   (D) 85

188. Let a and b be numbers such that $a^3 = b^2$. Which of the following is equivalent to $b\sqrt{a}$?

(A) $b^{\frac{2}{3}}$ (B) $b^{\frac{4}{3}}$ (C) b^2 (D) b^3 (E) b^4

189. Let m and n be positive integers such that one-third of m is n less than one-half of m. Which of the following is a possible value of m?

(A) 15   (B) 21   (C) 24   (D) 26   (E) 28

190. If a and b are numbers such that $(a-4)(b+6)=0$, then what is the smallest possible value of $a^2 + b^2$?

191. Clarissa and Shawna, working together, can paint the exterior of a house in 6 days. Clarissa by herself can complete this job in 5 days less than Shawna. How long will it take Clarissa to complete the job by herself?

(A) 16 days   (B) 17.5 days   (C) 19 days   (D) 18.5 days

192. Find k such that $f(x) = x^3 - kx^2 + kx + 2$ has the factor $x - 2$.

(A) 5   (B) 6   (C) 7   (D) 8

193. Let $f(x) = ax^2$ and $g(x) = bx^4$ for any value of x. If a and b are positive constants, for how many values of x is $f(x) = g(x)$?

(A) None   (B) One   (C) Two   (D) Three   (E) Four

194. Let a and b be numbers such that $30 < a < 40$ and $50 < b < 70$. Which of the following represents all possible values of $a-b$?

(A) $-40 < a-b < -20$ (B) $-40 < a-b < -10$ (C) $-30 < a-b < -20$

(D) $-20 < a-b < -10$ (E) $-20 < a-b < 30$

195. Given that a, b, c are the roots of the equation $x^3 - 5x^2 - 7x + 14 = 0$, find $\frac{1}{a} + \frac{1}{b} + \frac{1}{c}$.

(A) $\frac{2}{7}$ (B) $\frac{5}{14}$ (C) $\frac{1}{2}$ (D) $\frac{3}{5}$

196. In the equation shown below, x, y and z are positive integers. All of the following could be a possible value of y EXCEPT

$$\frac{x}{3} + \frac{y}{12} = z$$

(A) 4   (B) 6   (C) 8   (D) 12   (E) 20

197. In the equation below, m and n are integers such that $m > n$. Which of the following is the value of m?

$$\sqrt{72} + \sqrt{72} = m\sqrt{n}$$

(A) 6   (B) 12   (C) 16   (D) 24   (E) 48

198. The table above shows some values for the function N. If $N(t) = k \cdot 2^{-at}$ at for

t	0	1	2
$N(t)$	128	16	2

positive constants k and a, what is the value of a ?

(A) -3 (B) -2 (C) $\dfrac{1}{3}$ (D) 2   (E) 3

199. Amy is two years older than Bill. The square of Amy's age in years is 36 greater than the square of Bill's age in years. What is the sum of Amy's age and Bill's age in years?

200. If $9^x - 9^{x-1} = 216$, then the value of 2^x is:

(A) $4\sqrt{2}$ (B) $12\sqrt{2}$ (C) $10\sqrt{5}$ (E) $4\sqrt{10}$

201. If $(x+1)^2 = 4$ and $(x-1)^2 = 16$, what is the value of x ?

(A) -3 (B) -1 (C) 1   (D) 3   (E) 5

202. The value of y increased by 12 is directly proportional to the value of x decreased by 6. If $y = 2$ when $x = 8$, what is the value of x when $y = 16$?

(A) 8   (B) 10   (C) 16   (D) 20   (E) 28

203. Maximize $z = 2x + y$ subject to $x \geq 0$, $y \geq 0$, $x + y \leq 6$, $x + y \geq 1$.

(A) 6   (B) 8   (C) 10   (D) 12

204. Two cars are racing at a constant speed around a circular racetrack. Car A requires 15 seconds to travel once around the racetrack, and car B requires 25 seconds to travel once around the racetrack. If car A passes car B, how many seconds will elapse before car A once again passes car B?

205. To measure the height of Lincoln's caricature on Mt. Rushmore, two sightings 800 feet from the base of the mountain are taken. If the angle of elevation to the bottom of Lincoln's face is $32°$ and the angle of elevation to the top is $35°$, what is the height of Lincoln's face accurate to two decimal places?

(A) 30.15 ft   (B) 36.29 ft   (C) 45.12 ft   (D) 60.27 ft

206. Let $x+3, 2x+1$, and $5x+2$ be consecutive terms of an arithmetic sequence. Find the absolute value of the common difference of the terms.

(A) 2   (B) $\dfrac{5}{2}$ (C) $\dfrac{7}{2}$ (D) 4

207. A ball is dropped from a height of 30 feet. Each time that it strikes the ground, it bounces up to 0.8 of the previous height. How many times does the ball need to strike the ground before its height remains less than 6 inches?

(A) 3   (B) 5   (C) 6   (D) 8

208. If $\dfrac{3}{x-3} + \dfrac{5}{2x-6} = \dfrac{11}{2}$, then the value of $2x-6$ is

(A) 2   (B) 12   (C) 6   (D) 8

209. At Springfield University, there are 10 000 students, and there are as many male students as female students. Each student is enrolled either in the Arts program or Science program (but not in both); 60% of the students are in the Arts program. Also, 40% of the Science students are male. To the nearest percent, what percentage of the Arts students are female?

(A) 50%   (B) 52%   (C) 26%   (D) 43%

210. What is the largest integer n for which $3(n^{2007}) < 3^{4014}$?

(A) 2   (B) 3   (C) 6   (D) 8

211. There are a certain number of red balls, green balls and blue balls in a bag. Of the balls in the bag, $\frac{1}{3}$ are red and $\frac{2}{7}$ are blue. The number of green balls in the bag is 8 less than twice the number of blue balls. The number of green balls in the bag is

(A) 12   (B) 16   (C) 20   (D) 24

212. In a survey of 270 college students, it is found that 64 like cabbage, 94 like broccoli, 58 like cauliflower, 26 like both cabbage and broccoli, 28 like both cabbage and cauliflower, 22 like both broccoli and cauliflower, and 14 like all three vegetables. How many of the 270 students do not like any of these vegetables?

(A) 96   (B) 116   (C) 132   (D) 140

213. An urn contains 7 white balls and 3 red balls. Three balls are selected. In how many ways can the 3 balls be drawn from the total of 10 balls if 2 balls are white and 1 is red?

(A) 45   (B) 56   (C) 63   (D) 84

214. Five positive integers are listed in increasing order. The difference between any two consecutive numbers in the list is three. The fifth number is a multiple of the first number. How many different such lists of five integers are there?

(A) 3   (B) 4   (C) 5   (D) 6   (E) 7

215. In 2004, Gerry downloaded 200 songs. In 2005, Gerry downloaded 360 songs at a cost per song which was 32 cents less than in 2004. Gerry's total cost each year was the same. The cost of downloading the 360 songs in 2005 was

(A) $144.00   (B) $108.00   (C) $80.00   (D) $259.20   (E) $72.00

216. If w is a positive integer and $w^3 = 9w$, then w^5 is equal to

(A) 59049   (B) 243   (C) 1024   (D) 3125   (E) 32

217. If p, q and r are positive integers and $p + \dfrac{1}{q + \dfrac{1}{r}} = \dfrac{25}{19}$, then q equals

(A) 1   (B) 2   (C) 3   (D) 4

218. If $a = 7$ and $b = 13$, the number of even positive integers less than ab is

(A) $\dfrac{ab-1}{2}$ (B) ab (C) $ab-1$ (D) $\dfrac{a+b}{4}$

219. In the board game "Silly Bills", there are $1, $2 and $3 bills. There are 11 more $2 bills than $1 bills. There are 18 fewer $3 bills than $1 bills. If there is $100 in total, then how many $1 bills are there in the board game?

(A) 11   (B) 14   (C) 22   (D) 33

220. A box contains apple and pears. An equal number of apples and pears are rotten. $\frac{2}{3}$ of all of the apples are rotten. $\frac{3}{4}$ of all of the pears are rotten. What fraction of the total number of pieces of fruit in the box is rotten?

(A) $\frac{17}{24}$ (B) $\frac{7}{12}$ (C) $\frac{5}{8}$ (D) $\frac{12}{17}$

221. If x and y are integers with $(y-1)^{x+y} = 4^3$, then the number of possible values for x is

(A) 8   (B) 3   (C) 4   (D) 6

222. Simplify: $(x+2)(x^2+2x+3)$

(A) x^2+7x+6 (B) $5x^2+7x+6$ (C) $2x^3+x^2+x+6$ (D) x^3+4x^2+7x+6

223. What are all values of x for which $4-x^2 \geq x-2$

(A) $x \geq -3$ (B) $-5 \leq x \leq 0$ (C) $-3 \leq x \leq 2$ (D) $x \leq -3$ or $x \geq 2$

224. Which binomial is a factor of $3x^2+2x-5$?

(A) $3x-1$ (B) $x-1$ (C) $3x-5$ (D) $x-5$

225. Simplify: $\dfrac{14c^3d^2 - 21c^2d^3}{14cd^2}$?

(A) $c^2 - \dfrac{3cd}{2}$ (B) $c^2 - \dfrac{3c^2d}{2}$ (C) $c^2 - 21c^2d^2$ (D) $c^2d - \dfrac{3cd}{2}$

226. To find the image length, L , of a 4-foot-tall object in a spherical mirror with a focal length of 2 feet, $L = 4(\dfrac{2}{a-2})^2$ can be used, where a is the distance, in feet, of the object from the mirror. What is the image length of the object when it is 1.5 feet away from the mirror?

(A) 256 feet   (B) 128 feet   (C) 64 feet   (D) 32 feet

227. The number of bacteria in a culture can be represented by the formula $N_t = 2.5 N_{t-1}$. In the formula, N_t is the number of bacteria at the end of t minutes, and N_{t-1} is the number of bacteria at the end of $t-1$ minutes. There are 16,406 bacteria in the culture at the end of 7 minutes. How many bacteria will be in the culture at the end of 10 minutes?

(A) 23,437   (B) 102,538   (C) 123,045   (D) 256,343

228. The distance required for a car to stop is directly proportional to the square of its velocity. If a car can stop in 112.5 meters at 15 kilometers per hour, how many meters are needed to stop at 25 kilometers per hour?

(A) 250.75   (B) 298.00   (C) 312.50   (D) 337.50

229. Danny bought a car for $15,000 and its value depreciated linearly. After 3 years the value was $11,250. What is the amount of yearly depreciation?

(A) $2,000   (B) $1,500   (C) $1,250   (D) $750

230. In 1994, the average price of a new domestic car was \$16,930. In 2002, the average price was \$19,126. Based on a linear model, what is the predicted average price for 2008?

(A) \$22,969   (B) \$21,322   (C) \$20,773   (D) \$18,577

231. If the graph of a line has a positive slope and a negative y-intercept, what happens to the x-intercept if the slope and the y-intercept are doubled?

(A) The x-intercept becomes four times larger.
(B) The x-intercept becomes twice as large.
(C) The x-intercept becomes one-fourth as large.
(D) The x-intercept remains the same.

232. An object is fired upward at an initial velocity, v_o, of 240 ft/s. The height, $h(t)$,of the object is a function of time, t , in seconds and is given by the formula $h(t) = v_o t - 16t^2$. How long will it take the object to hit the ground after takeoff?

(A) 16 seconds   (B) 15 seconds   (C) 7.5 seconds   (D) 4 seconds

233. Give values for A, B and C so that $-3x^2 + 7x + 11 = A(x+B)^2 + C$

234. Given $f(x) = -3x^2 + 5$, what is the range of the function?

(A) all real numbers less than or equal to 5
(B) all integers less than or equal to 5
(C) all nonnegative real numbers   (D) all nonnegative integers

235. A store received $823 from the sale of 5 tape recorders and 7 radios. If the receipts from the tape recorders exceeded the receipts from the radios by $137, what is the price of a tape recorder?

(A) $49   (B) $68   (C) $84   (D) $96

236. A region is defined by this system: $y > 2x + 1$
$$y \leq -x + 2$$
In which quadrants of the coordinate plane is the region located?

(A) I, II, III only   (B) II, III only   (C) III, IV only   (D) I, II, III, IV

237. When Robert was born, his grandfather invested $1,000 for his college education. At an interest rate of 4.5%, compounded annually, **approximately** how much would Robert have at age 18? (use the formula $A = P(1 + r)^t$ where P is the principal, r is the interest rate, and t is the time in years)

(A) $1,810   (B) $2,200   (C) $3,680   (D) $18,810

238. A new automobile is purchased for $20,000. If $V = 20000(0.8)^x$ gives the car's value after x years, **about** how long will it take for the car to be worth half its purchase price?

(A) 3 years   (B) 4 years   (C) 5 years   (D) 6 years

239. Factor completely: $18x^4 - 32x^2$

(A) $2(3x^2 - 4x)(3x^2 + 4x)$ (B) $2x^2(9x^2 - 16)$ (C) $2x^2(3x - 4)^2$
(D) $2x^2(3x - 4)(3x + 4)$

240. What is the solutions of $|2x-3|>7$?

241. Which of the following expression is equivalent to $(\dfrac{2 \cdot x^{-2} \cdot y^{3}}{z^{-1}})^{-3}$?

(A) $\dfrac{8x^{6}y^{-3}}{z^{3}}$ 　　(B) $\dfrac{8x^{6}z^{3}}{8y^{9}}$ 　　(C) $\dfrac{8x^{6}}{z^{3}y^{6}}$ 　　(D) $\dfrac{x^{6}}{8y^{9}z^{3}}$

242. At how many points do the curves $y=x$ and $y=x^{2003}$ intersect?

(A) three 　　(B) two 　　(C) one 　　(D) zero

243. What can you say about the intersection of lines $4x-y=7$ and $x+3y=5$?

(A) They intersect at a point (x, y) ; both x and y are positive.

(B) They intersect at a point (x, y) ; both x and y are negative.

(C) They intersect at a point (x, y) ; x is positive and y is negative.

(D) They intersect at a point (x, y) ; x is negative and y is positive.

244. Solve the equation $\sqrt{x+4}=x$. Which of the following statements describes the solution set?

(A) There is one solution; it is a rational number.

(B) There are two solutions; they are rational numbers.

(C) There is one solution; it is an irrational number.

(D) There are two solutions; they are irrational numbers.

245. Solve the quadratic equation $2x^2 + 7x + 5 = 0$. What is the sum of the two solutions?

(A) -7 (B) $-\dfrac{3}{2}$ (C) $-\dfrac{7}{2}$ (D) $\dfrac{7}{2}$

246. Find the two solutions to the quadratic equation $x^2 - x = 2$. What is the larger of the two solutions?

(A) 2   (B) 1   (C) 0   (D) -1

247. You are selling tickets to a concert priced at $5 for adults and $2 for students. At the end of the day, you have sold 10 tickets and you report that you have taken in $40. However, you may have made a $1 error, and so your correct total may have been $39 or $40 or $41. Can the number of student tickets sold be determined?

(A) Yes, 3 were sold.   (B) Yes, 4 were sold.   (C) Yes, 5 were sold.
(D) Yes, none were sold.

248. Solve the inequalities $-6 < 2x + 4 \leq 0$ for x . How many integers are in the solution set?

(A) 3   (B) 4   (C) 6   (D) 7

249. Solve the equation $q = 11 - 4p$ for p .

(A) $p = 4q + 11$ (B) $p = 4q - 11$ (C) $p = \dfrac{q + 11}{4}$ (D) $p = \dfrac{11 - q}{4}$

250. Compare $\sqrt{m+n}$ and $\sqrt{m}+\sqrt{n}$, where m and n are known to be positive numbers between 10 and 99, but their actual values are unknown. Which of the following statements must be true?

(A) $\sqrt{m+n}$ and $\sqrt{m}+\sqrt{n}$ are always equal
(B) $\sqrt{m+n}$ is always larger than $\sqrt{m}+\sqrt{n}$.
(C) $\sqrt{m+n}$ is always smaller than $\sqrt{m}+\sqrt{n}$.
(D) $\sqrt{m+n}$ is usually larger than $\sqrt{m}+\sqrt{n}$.

251. What is the LCM of $4ab$ and $6bc^2$?

(A) $2ab$ (B) $2b$ (C) $12abc^2$ (D) $12ab^2c^2$

252. A band wants to distribute its music on CD's. The equipment to produce the CD's costs $250, and blank CD's cost $5.90 for a package of 10. Which of the following represents the total cost, in dollars, to produce n CD's, where n is a multiple of 10?

(A) $(250+0.59)n$ (B) $250+0.59n$ (C) $(250+5.90)n$ (D) $250n+5.90$

253. Kim' Fruit Market prices fruit by the piece and not by weight. A small bag of three apples and two pears costs $1.85. A large bag of seven apples and five pears costs $4.45. Which of the following statements must be true?

(A) An apple and a pear are the same price.
(B) A pear is 5 cents more than an apple.
(C) A pear is 10 cents more than an apple.
(D) An apple is 5 cents more than a pear.

254. If $f(x)=x^4-3x^3-9x^2+4$, for how many real numbers k does $f(k)=2$?

(A) One   (B) Two   (C) Three   (D) Four

255. If $f(x)=x+3$ and $g(x)=\dfrac{x^2-9}{x-3}$, which of the following statements are true about the graphs of f and g in the xy-plane?

 I. The graphs are exactly the same   II. The graphs are the same except when $x=3$.
 III. The graphs have an infinite number of points in common.

(A) I only   (B) II only   (C) I and III   (D) II and III

256. If line l is the perpendicular bisector of the line segment with endpoins (2, 0) and (0, −2), what is the slope of line l

(A) 2   (B) 1   (C) 0   (D) −1

257. Simplify $(a-b)^3+(a+b)^3$

(A) a^3+b^3 (B) $2a^3+3ab^2$ (C) $2a^3+6a^2b+6ab^2+2b^3$ (D) none of these

258. Chris is thinking of two numbers whose sum is 9 and whose difference is 3. What is the product of these two numbers?

(A) 6   (B) 12   (C)15   (D) 18

259. Simplify $(3^{-1}+2^{-1})^{-1}$

(A) 5   (B) $\dfrac{5}{6}$ (C) $\dfrac{6}{5}$ (D) $\dfrac{1}{5}$

260. If the measure of one angle of a rhombus is $60°$, then the ratio of the length of its longer diameter to the length of its shorter diagonal is

(A) 2   (B) $\sqrt{3}$ (C) $\sqrt{2}$ (D) $\dfrac{\sqrt{3}}{2}$

261. A number of n is increased by 8. If the cube root of that result equals -0.5, what is the value of n?

(A) -15.625 (B) -8.794 (C) -8.125 (D) -7.875

262. If a and b are real numbers, $i^2 = -1$ and $(a+b) + 5i = 9 + ai$, what is the value of b?

(A) 4   (B) 5   (C) 9   (D) $4+5i$

263. Twenty students have each sampled one or more of three kinds of candy bars that a school store sells. If 3 students have sampled all three kinds, and 5 have sampled exactly two kinds, how many of these students have sampled only one kind?

(A) 8   (B) 12   (C) 15   (D) 17

264. $\triangle ABC$ has a right angle at C. If the length of side AC is 10 and the measure of $\angle BAC$ is $22°$, what is the length of side BC?

(A) 3.7   (B) 4.0   (C) 5.8   (D) 6.8

265. Solve the formula $E = mc^2$ for c.

(A) $c = \sqrt{Em}$ (B) $c = \dfrac{\sqrt{Em}}{m}$ (C) $c = \dfrac{E}{mc}$ (D) $c = \dfrac{\sqrt{E}}{m}$

267. The expression $(2-x)^3$ is equivalent to which of the following?

(A) $8 - 4x - 2x^2 - x^3$ (B) $8 - 4x + 2x^2 - x^3$ (C) $8 + 4x - 2x^2 - x^3$
(D) $8 - 12x + 6x^2 - x^3$

268. The function h given by $h(t) = -16t^2 + 46t + 5$ represents the height of a ball, in feet, t seconds after it is thrown. To the nearest foot, what is the maximum height the ball reaches?

(A) 5   (B) 23   (C) 35   (D) 38

269. Find all solutions of $\dfrac{3x^2}{x+5} = \dfrac{x^2 - 9x + 5}{x+5}$.

(A) -5 (B) $-\dfrac{1}{2}$ (C) $\dfrac{1}{2}$ (D) -5 and $\dfrac{1}{2}$

270. The front, side, and bottom faces of a rectangular solid have areas of 24 square centimeters, 8 square centimeters, and 3 square centimeters, respectively. What is the volume of the solid, in cubic centimeters?

(A) 24   (B) 96   (C) 192   (D) 288

271. $(a + a^{-1})^{-1} =$

(A) $a^{-1} + a$ (B) $\dfrac{a}{a^2 + 1}$ (C) $\dfrac{a^2 + 1}{a}$ (D) $\dfrac{a}{a+1}$

272. If $2a = b, 3b = c,$ and $a + c = 70,$ find $a + b + c.$

(A) 30   (B) 60   (C) 70   (D) 90

273. If $x - 2$ is a factor of $x^3 + kx^2 + 12x - 8,$ then $k =$

(A) -6 (B) -3 (C) 2   (D) 3

274. In a group of 10 people, 60 percent have brown eyes. Two people are to be selected at random from the group. What is the probability that neither person selected will have brown eyes?

(A) 0.13   (B) 0.16   (C) 0.25   (D) 0.36

275. What is the range of the function defined by $f(x) = \dfrac{1}{x} + 2$

(A) all real numbers   (B) all real numbers except $-\dfrac{1}{2}$

(C) all real numbers except 0   (D) all real numbers except 2

276. If $(a+b)^{\frac{1}{2}} = (a+b)^{-\frac{1}{2}}$, which of the following must be true?

(A) $b = 0$ (B) $a + b = 1$ (C) $a - b = 1$ (D) $a^2 + b^2 = 1$ (E) $a^2 - b^2 = 1$

277. If $\sin\theta = 0.57$, then $\sin(\pi - \theta) =$

(A) -0.57 (B) -0.43 (C) 0.43 (D) -0.57

278. Find the sum of the roots of $ax^2 + bx + c = 0$.

(A) $\dfrac{b}{a}$ (B) $\dfrac{c}{a}$ (C) $-\dfrac{b}{a}$ (D) $-\dfrac{c}{a}$ (E) $\dfrac{c}{b}$

279. Find the product of the roots of $ax^2 + bx + c = 0$.

(A) $\dfrac{b}{a}$ (B) $\dfrac{c}{a}$ (C) $-\dfrac{b}{a}$ (D) $-\dfrac{c}{a}$ (E) $\dfrac{c}{b}$

280. What is the measure of one of the larger angles of a parallelogram in the xy-plane that has vertices with coordinates (2, 1), (5, 1), (3, 5), and (6,5)?

(A) 93.4 °   (B) 96.8 °   (C) 104.0 °   (D) 108.3 °

281. If $f(x) = \sqrt[3]{x^3 + 1}$, what is $f^{-1}(1.5)$?

(A) 2.4   (B) 1.6   (C) 1.5   (D) 1.3

282. If $f(x)$ is a function for which $f(2+k) = f(2-k)$ and $f(-3) = 0$, for which of the following is $f(x)$ also equal to zero?

(A) $x = 3$ (B) $x = 5$ (C) $x = -1$ (D) $x = 6$ (E) $x = 7$

283. If $f(x, y) = \dfrac{1}{4}x - y$, then which of the following is equal to $f(8, 3)$?

(A) $f(12, 2)$ (B) $f(16, 6)$ (C) $f(2, 1)$ (D) $f(-12, -2)$

284. Assuming $a \neq 0$, $\dfrac{5 - \dfrac{1}{a}}{a^{-1}} =$

(A) $5a - 1$ (B) $\dfrac{5a - 1}{a^2}$ (C) 4   (D) $\dfrac{1 - 5a}{a^2}$

285. If $\sec\theta = 2$, then $\cos\theta \sec\theta =$

(A) 2   (B) 4   (C) $\dfrac{1}{4}$ (D) 1

286. If $16x^4 - 9 = 4$, then x could equal which of the following?

(A) 1.34   (B) -0.95 (C) 0.87   (D) 0.90

287. What is the range of $f(x) = \sqrt{4 - x^2}$?

(A) $y \geq 0$ (B) $y \geq 2$ (C) $-2 \leq y \leq 2$ (D) $0 \leq y \leq 2$

288. Simplify the expression $3 + 5(9 - 6)^2 \div 3 - 2$.

(A) 12   (B) 16   (C) 17   (D) 48   (E) 136

289. Solve the following for x : $\dfrac{5}{3}(x - 3) = \dfrac{3}{2}(x - 2)$

(A) 14   (B) 12   (C) 3   (D) $\dfrac{13}{2}$ (E) 1

290. Which of the following is a polynomial with roots 0, 4, and i?

(A) $x^3 - 4x^2 + x - 4$ (B) $x^4 - 4x^2 + x - 4$ (C) $x^4 - 4x^3 + x^2 - 4x$
(D) $x^4 - 4x^3 - x^2 + 4x$

291. If θ is an acute angle and $\sin\theta = \dfrac{3}{4}$, then $\cos 2\theta =$

(A) $-\dfrac{1}{8}$ (B) $-\dfrac{7}{25}$ (C) 1   (D) -2

292. Find all real x so that the following statements is true: $\dfrac{x}{2}-1 \le 1-\dfrac{x}{2}$.

(A) $x \le -2$ (B) $x \le 0$ (C) $x \le 2$ (D) All real number

293. What are the asymtote(s) of $f(x)=\dfrac{6x^2}{4-x^2}$?

(A) $x=2$ and $x=-2$ (B) $y=-6$ (C) $x=2$
(D) $x=2$, $x=-2$, and $y=-6$

294. If $2a=b$, $3b=c$, and $a+c=70$, find $a+b+c$.

(A) 30   (B) 60   (C) 70   (D) 90

295. If $3^{5-x}=81^{x+1}$, what does x equal?

(A) $\dfrac{1}{3}$ (B) 3   (C) $\dfrac{4}{5}$ (D) $\dfrac{1}{5}$

296. The sum of twice one number plus three times the second is 7. Their sum is $2\dfrac{1}{2}$. Find the product of the numbers.

(A) -114 (B) -9 (C) 1   (D) $\dfrac{3}{2}$

297. If $f(x)=2x+5$ and $g(x)=\dfrac{1}{6+x}$, then $f(g(12))=$

(A) 29   (B) $\dfrac{1}{18}$ (C) $\dfrac{29}{18}$ (D) 47

298. Find the value of k so that the slope of the line passing through the points $(-2, 5)$ and $(4, k)$ is $-\dfrac{3}{4}$.

(A) $\dfrac{1}{2}$ (B) 2   (C) -3 (D) $-\dfrac{13}{4}$

299. Seven integers are arranged from least to greatest. If the median is 9 and the only mode is 7, what is the least possible range for the 7 numbers?

(A) 4   (B) 5   (C) 6   (D) 8

300. In the simplest form, $\dfrac{x}{x-1} - \dfrac{2}{1-x} =$

(A) $\dfrac{x+2}{x-1}$ (B) $\dfrac{x-2}{x-1}$ (C) $\dfrac{x-2}{x^2-1}$ (D) $x+2$ (E) -2

301. The graph of $x^2 - xy = 4$ has which of the following symmetries?

(A) Symmetric with respect to the x-axis   (B) Symmetric with respect to the y-axis
(C) Symmetric with respect to the origin   (D) Symmetric with respect to both axes

302. Solve for x: $2(x-5)(x+3) = -28$

(A) $\{-2, 5\}$ (B) $\{-3, -14\}$ (C) $\{\dfrac{4 \pm \sqrt{3}}{2}\}$ (D) $\{1 \pm \sqrt{2}\}$

303. Find x if $\dfrac{x}{12} - \dfrac{x+2}{4} < 0$.

(A) $x < -3$ (B) $x > -3$ (C) $x < 3$ (D) $x > 3$

304. Which of the following is not equivalent to i^{21}?

(A) i^{17} (B) i^{9} (C) i^{31} (D) i^{105}

305. The value of $(3^{-1}-2^{-1})^{-1}$ is

(A) -6 (B) -1 (C) $-\dfrac{1}{6}$ (D) 1

306. Cost is a function of the number of units produced as given by: $C(n)=0.01n^2-90n+25,000$. How many units, n, produce a minimum cost C?

(A) 500   (B) 4500   (C) 9000   (D) 18000

307. In simplest $-3[4x-2\{3x-4(x-1)-2\}-3]+7=$

(A) $-18x+20$ (B) $-6x+4$ (C) $-18x+28$ (D) $-6x+52$

308. $\sqrt[5]{\sqrt[4]{\sqrt[3]{\sqrt{n}}}}=$

(A) $n^{\frac{1}{17}}$ (B) $n^{\frac{1}{19}}$ (C) $n^{\frac{1}{60}}$ (D) $n^{\frac{1}{120}}$

309. The solution of $\begin{cases} 3x+4y \geq 24 \\ 5x-3y \geq 15 \end{cases}$ lies in quadrants

(A) I only   (B) I and II   (C) I and III   (D) I and IV

310. If $x+2$ is a factor of $x^4+x^3+3x^2+kx-10$, then $k=$

(A) -13 (B) 15   (C) 13   (D) 5

311. Which of the following is a factor of: $\dfrac{a^2}{16} - \dfrac{4b^2}{9}$?

(A) $\dfrac{a}{4} + \dfrac{2b}{9}$ (B) $\dfrac{a}{4} - \dfrac{2b}{9}$ (C) $\dfrac{a}{4} - \dfrac{2b}{3}$ (D) $\dfrac{a}{2} - \dfrac{2b}{3}$

312. Which of the following is the equation of a line with x-intercept $(6, 0)$ and y-intercept $(0, -15)$?

(A) $y = \dfrac{5}{2}x - 15$ (B) $y = -\dfrac{5}{3}x - 15$ (C) $y = -\dfrac{5}{2}x - 6$ (D) $y = -\dfrac{2}{5}x + 15$

313. If your average on the first three mathematics tests is 82, what must your score on fourth test be so that your overall average is 86?

(A) 92   (B) 94   (C) 96   (D) 98

314. Which of the following describes the right and left behavior of the graph of $f(x) = -3x^7 + 2x^5 - 3x + 6$?

(A) Rises right and left   (B) Falls right and left   (C) Falls left, rises right
(D) Rises left, falls right

315. The simplified form of the expression $(x+3)^2 - (x-3)^2$ is

(A) 9   (B) 18   (C) $12x$ (D) $2x^2 + 18$ (E) $18x$

316. The radius of a circle is 8 cm less than its diameter.
What is the area of the circle, in centimeters?

(A) 8π
(B) 16π
(C) 36π
(D) 64π

318. Factor $a^3 - a^2b - ab^2 + b^3$ completely.

(A) $(a^2 - b^2)(a+b)$ (B) $(a^2 + b^2)(a+b)$ (C) $(a-b)(a+b)^2$ (D) $(a-b)^2(a+b)$

319. Line l passes through the origin and is parallel to the line $y = \dfrac{2}{3}x - 6$. What is the sum of the coordinates of the point where the line l intersects the line $y = \dfrac{1}{2}x - 4$?

(A) -40 (B) -10 (C) 0 (D) 12

320. The operation Φ is defined for all real numbers a and b by the operation: $a\Phi b = a^{-b} - 3b$. If $n\Phi - 2 = 70$, which of the following could equal n?

(A) 7   (B) 9   (C) -8 (D) 8.7

321. Simplify $\sqrt[3]{64x^3y^{27}}$.

(A) $21xy^9$ (B) $8xy^3$ (C) $4xy^3$ (D) $4xy^9$

322. Factor $12x^3 + 5x^2 - 72x$ completely.

(A) $x(4x-9)(3x+8)$ (B) $(4x^2-9)(3x+8)$ (C) $(4x+9)(3x^2-8)$
(D) $2x(2x-9)(3x+8)$ (E) $x(4x+9)(3x-8)$

323. The lines $x + 2y = 7$ and $2x + ky = 5$ are perpendicular if the value of k is

(A) -4 (B) -1 (C) 1 (D) 4

324. Solve for y in terms of x if $x = \dfrac{2}{1-2y}$.

(A) $\dfrac{2+x}{2x}$ (B) $\dfrac{2x-1}{2x}$ (C) $\dfrac{2-x}{2x}$ (D) $\dfrac{x-2}{2x}$

325. Mr. Lee left an estate to be divided among his wife and three sons. The first son is to receive $\dfrac{1}{6}$ of the total, the second $\dfrac{1}{12}$, the third $\dfrac{1}{4}$, and the wife \$40,000. What was the total value of the estate?

(A) \$40,000   (B) \$80,000   (C) \$120,000   (D) \$160,000

326. If $12\sin^2 x + \sin x - 1 = 0$ over the interval $180° \le x \le 360°$, then

(A) $-19.5°$ (B) $194.5°$ or $344.5°$ (C) $199.5°$ (D) $199.5°$ or $340.5°$

327. Daniel can repair a car alone 7 hours. Peter can repair the same car alone in 5 hours. How long would it take them to repair the car together?

(A) $2\dfrac{11}{12}$ hours   (B) 6 hours   (C) $3\dfrac{4}{7}$ hours   (D) $3\dfrac{1}{12}$ hours

328. How many solutions does the equation $|2x - 2| = x$ have?

(A) 1   (B) 2   (C) 3   (D) Infinite many

329. $\dfrac{\tan\theta + \cot\theta}{\tan\theta} =$

(1) 1   (B) $\csc^2\theta$ (C) $\sin^2\theta$ (D) $\sec^2\theta$

330. Given $z = x^2 + 3x + \dfrac{y^2 + 2xy}{x+2}$, find the value of z when $x = 2$.

(A) $\dfrac{y^2 + 4y + 10}{4}$ (B) $\dfrac{y^2 + 4y + 32}{2}$ (C) $\dfrac{y^2 + 4y + 40}{4}$ (D) $y^2 + y + 10$

331. If $\dfrac{m}{3m - n} = \dfrac{1}{5}$, then $\dfrac{n}{m} =$

(A) -2 (B) $-\dfrac{1}{2}$ (C) $\dfrac{1}{2}$ (D) 2

332. What is the reciprocal of $6 + i$?

(A) $\dfrac{1}{6}$ (B) $\dfrac{6 - i}{35}$ (C) $\dfrac{6 + i}{36}$ (D) $\dfrac{6 - i}{37}$

333. If $f(x) = \dfrac{4x^2 - 9}{2x + 3}$, what value does the function approach as x approaches $-\dfrac{3}{2}$?

(A) 0.02 (B) 0 (C) -4.33 (D) -6

334. The probability that it will snow tomorrow is $\dfrac{2}{3}$ and, independently, the probability that it will also snow the day after tomorrow is $\dfrac{1}{5}$. What is the probability that it will snow tomorrow but not the day after tomorrow?

(A) $\dfrac{2}{15}$ (B) $\dfrac{2}{5}$ (C) $\dfrac{3}{4}$ (D) $\dfrac{8}{15}$

335. In factored form, $x^3 - 2x^2 + 2x - 4 =$

(A) $(x-2)(x+2)^2$ (B) $(x-2)^2(x+2)$ (C) $(x-2)^3$ (D) $(x-2)(x^2+2)$

336. Solve the following for x: $-4 \leq -2(x+8) < 8$

(A) $0 < x \leq 6$ (B) $-12 < x$ or $x \leq -6$ (C) $-10 \leq x < 2$
(D) $-12 < x \leq -6$

337. Peter found a battery powered drill for 25% off the original price. At the checkout counter the clerk enters the sale price, adds 5% sales tax, and then tells Peter he owes $189. What was the original price of the drill?

(A) $158.78   (B) $198.45   (C) $240.00   (D) $226.80   (E) $302.40

338. $\sqrt[3]{20x^2y^7} \cdot \sqrt[3]{50x^4y^2} =$

(A) $20^{\frac{1}{3}}(50)^{\frac{1}{3}}x^6y^9$ (B) $70^{\frac{1}{3}}x^2y^3$ (C) $10x^{\frac{8}{3}}y^{\frac{14}{3}}$ (D) $10x^2y^3$

339. Solve for n if $4 \cdot 2^{n-1} = 8^n$.

(A) $\dfrac{1}{4}$ (B) $\dfrac{1}{2}$ (C) $\dfrac{5}{2}$ (D) 32   (E) No solution

340. For how many values of k does the equation $kx^2 + 2kx + 1 = 0$ has exactly one solution?

(A) 1   (B) 2   (C) 3   (D) 4   (E) none

341. If $\dfrac{x+2}{x-5} \geq 0$, then which of the following describes x?

(A) $x \geq -2$ (B) $-2 \leq x < 5$ (C) $x \leq -2$ or $x > 5$ (D) $x > 5$

342. Simplify: $5x-3[4y-2(x-3y)]$
(A) $11x+6y$ (B) $11x-30y$ (C) $-x-30y$ (D) $-x+6y$

343. Two students started walking from the same point in opposite directions for 12 feet each. Each then made a 90 degree turn to the left and right and walked another 5feet. How far apart are two students?

(A) 13 ft   (B) 24 ft   (C) 34 ft   (D) 26 ft

344. If $f(x)=2x+1$ then $\dfrac{f(x+h)-f(x)}{h} =$

(A) $\dfrac{2}{h}$ (B) 2   (C) $\dfrac{h-1}{h}$ (D) $\dfrac{h+2}{h}$

345. Which of the following expressions represents the statement " N varies jointly with the square of x and with y" ?

(A) $N=kx^2 y$ (B) $N=\dfrac{k}{y\sqrt{x}}$ (C) $N=ky\sqrt{x}$ (D) $N=kx^2 y^2$

346. If $x \neq 0$, $y \neq 0$, then $\dfrac{x^{-1}-y^{-1}}{x-y} =$

(A) $-\dfrac{1}{xy}$ (B) $\dfrac{1}{(x-y)^2}$ (C) $\dfrac{y^2-x^2}{xy}$ (D) $\dfrac{x^2-y^2}{xy}$

347. Solve the equation: $x + \sqrt{x-4} = 4$

(A) 5   (B) 4   (C) 0 and 4   (D) 4 and 5

348. A problem from the Rhind papyrus (1650 B.C.) states: A quantity and its $\frac{2}{3}$, its $\frac{1}{2}$, and its $\frac{1}{7}$ added together becomes 388. What is this quantity?

(A) 84   (B) 168   (C) 42   (D) 126

349. Solve the inequality: $|7 - 3x| > 2$

(A) $x < \frac{5}{3}$ or $x > 3$ (B) $\frac{5}{3} < x < 3$ (C) $x < \frac{5}{3}$ (D) $x > -3$

350. Simplify the expression $\frac{a}{b} - \frac{b}{a}$.

(A) -1 (B) $a - b$ (C) $\frac{a^2 - b^2}{ab}$ (D) $\frac{b^2 - a^2}{ba}$

351. What does the angle $-\frac{5\pi}{12}$ equal in degree measure?

(A) $-75°$ (B) $-15°$ (C) $-150°$ (D) $37.5°$

351. If $\cos 2\theta = \frac{3}{4}$, then $\dfrac{1}{\cos^2\theta - \sin^2\theta} =$

(A) -1 (B) $\frac{3}{4}$ (C) $\frac{4}{3}$ (D) 4

352. If $\cot\theta = 1.33$, then $\sin\theta =$

(A) 0.01   (B) 0.63   (C) 0.60   (D) 0.75

353. The complete factorization of $6x^4 + 40x^3 - 14x^2$ is

(A) $(6x^2 - 2x)(x^2 + 7x)$ (B) $(3x^2 - x)(2x^2 + 14x)$ (C) $2x^2(3x - 1)(x + 7)$

(D) $2x^2(3x + 1)(x - 7)$ (E) $x(3x - 1)(2x^2 + 14x)$

354. Peter wanted to purchase 2 dozen pencils and a pen. Those items cost $8.45 and he did not have enough money. So he decided to purchase 8 fewer pencils and paid $6.05. How much did the pen cost?

(A) $1.25   (B) $0.30   (C) $1.65   (D) $1.15

355. Solve $A = \dfrac{xy}{x+y}$ for x.

(A) $x = \dfrac{Ay}{A+y}$ (B) $x = \dfrac{y-A}{Ay}$ (C) $x = \dfrac{y}{A+y}$ (D) $x = \dfrac{-Ay}{A-y}$

356. What are the zeros of $f(x) = 8x^3 - 2x^2 - 3x$?

(A) $-\dfrac{1}{2}, \dfrac{3}{4}$ (B) $\dfrac{1}{2}, -\dfrac{3}{4}$ (C) $-\dfrac{1}{2}, \dfrac{3}{4}, \dfrac{1}{2}$ (D) $-\dfrac{1}{2}, \dfrac{3}{4}, 0$

357. If $3x^4 - 2x^2 - 3$ is divided by $x + 2$ then the remainder is

(A) 37   (B) 53   (C) 59   (D) -59

358. Solve $(3^x)^2 = 27^{x-1}$ for x.

(A) $\dfrac{3}{2}$ (B) 3   (C) $\dfrac{5}{2}$ (D) 2

359. If $a = 1$, $b = 2$ and $c = 3$, then the value of $\dfrac{a^b + b^c}{a + b}$ is

(A) 5   (B) $\dfrac{8}{3}$ (C) 3   (D) $\dfrac{10}{3}$

360. An operation is defined on any three real numbers by $a \triangle b \triangle c = a(c - b)$. If $2 \triangle 1 \triangle x = 10$, then $x =$

(A) -6 (B) -4 (C) 5   (D) 6

361. You were supposed to add A and B. But by accident, you subtracted B from A and got 4. This number is different from the correct answer by 12. What is A?

(A) 4   (B) 5   (C) 6   (D) 10

362. If i is a zero of the polynomial $p(x)$, then which of the following must be a factor of $p(x)$?

(A) x^2 (B) $x^2 + 1$ (C) $x^2 - 1$ (D) $x^2 - 2ix + 1$

363. Simplify: $\dfrac{\dfrac{1}{x-2} - \dfrac{1}{x+2}}{\dfrac{2}{x+2} + \dfrac{2}{x-2}}$

(A) x (B) $\dfrac{4x}{x+2}$ (C) $\dfrac{1}{x}$ (D) $\dfrac{4x}{x-2}$

364. For all $x \neq 2$, $f(x) = (x-2)(2-x)^{-1}$. Which of the must be a true statement?

    I. $f(1) = f(-1)$ II. $f(4) = f(0)$ III. $f(\frac{1}{2}) = f(-2)$

(A) I only   (B) I and II only   (C) II and III only   (D) I, II, and III

365. Simplify: $\dfrac{7 + 20x - 3x^2}{2x^2 - 11x - 21}$

(A) $-\dfrac{3x+1}{2x+3}$ (B) $\dfrac{3x+1}{2x+3}$ (C) $\dfrac{1}{2}(3x+1)$ (D) $-2(x-7)$

366. If x and y are any real numbers, then which of the following statement(s) is/are always true?

    I. $(xy^2)^3 = x^3 y^6$ II. $\sqrt{-x^9} = \dfrac{1}{x^3}$ III. $(x^3 \times x^2)^2 = x^{12}$

(A) I only   (B) II only   (C) I and II   (D) I and III   (E) I, II, and III

367. A pair of pants and a sweater cost $98. If the sweater costs $16 more than the pants, how much did the sweater cost?

(A) $57   (B) $41   (C) $82   (D) $114

368. Solve the formula $S = \dfrac{11(n-2)}{5}$ for n:

(A) $5S - 9$ (B) $\dfrac{5(S+2)}{11}$ (C) $\dfrac{5S}{11} + 2$ (D) $\dfrac{5S+2}{11}$

369. It is 7 miles between Peter's house and Daniel's house. Peter and daniel started from their own houses, walking to each other's houses, at noon, and then meeting at 1:30 pm. Peter walked 1 mile per hour faster than Daniel. How fast did Daniel walk?

(A) $3\frac{1}{3}$ mph   (B) $2\frac{2}{3}$ mph   (C) $1\frac{5}{6}$ mph   (D) $2\frac{3}{7}$ mph

370. If $f(x)=5x-1$ and $g(f(-3))=18$, then $g(x)$ could be which of the following?

(A) $x+2$ (B) $-x+2$ (C) $-2x-12$ (D) $\frac{x}{4}+14$

371. Solve for x: $\sqrt{x+3}+3=4$

(A) 4   (B) -2 (C) 46   (D) 2

372. Given the expression $\dfrac{x^2-21}{x(x+2)}$, determine the domain of the variable x.

(A) $\{x\neq 0,\,-2\}$ (B) $\{x\neq -2\}$ (C) $\{x\neq 0\}$ (D) $\{x=0\}$

373. A car is purchased for \$24,500. If it decreases in value at a rate of 7% per year, how much will it be worth in 5 years?

(A) \$15,851   (B) \$15,925   (C) \$17,044   (D) \$22,785

374. Find x such that $\dfrac{2x}{3}-4\leq -1+\dfrac{2x}{3}$.

(A) $x\leq -\dfrac{15}{4}$ (B) $x\geq -\dfrac{9}{4}$ (C) $x\leq \dfrac{9}{4}$ (D) All real number

375. For how many different integral values of b are both roots of $x^2 + bx - 16 = 0$ integers?

(A) 3   (B) 4   (C) 5   (D) 6

376. Which of the following is true about the line whose equation is $4x - 2y - 10 = 0$?

(A) The x-intercept is 4 and the y-intercept is -2.

(B) The x-intercept is $\dfrac{5}{2}$ and the y-intercept is 5.

(C) The x-intercept is $\dfrac{5}{2}$ and the y-intercept is -5.

(D) The x-intercept is 5 and the y-intercept is $\dfrac{5}{2}$.

377. Five-sixths of the students in class are passing with a grade of C-or better. Three-fourths of the students in the same class are passing with a grade of B-or better. What fraction of the class is not passing with a grade of B-or better but is passing with a grade of C-or better?

(A) $\dfrac{5}{8}$ (B) $\dfrac{1}{12}$ (C) $\dfrac{1}{6}$ (D) $\dfrac{1}{4}$

378. What is the range of the piecewise function: $f(x) = \begin{cases} x^2 & x > 3 \\ -\dfrac{2}{3}x + 11 & x \leq 3 \end{cases}$?

(A) All real number   (B) $y \geq 0$ (C) $-3 \leq y \leq 3$ (D) $y \geq 9$

379. If the sum of the squares of two number is equal to the square of their sum, then the product of these two numbers must be

(A) 0   (B) 1   (C) 4   (D) 16

380. Solve the following for x: $|x-3|=-2$

(A) $x=1$ only   (B) $x=5$ only   (C) $x=1$ or $x=5$ (D) No solution

381. A classroom contained an equal number of boys and girls. Eight girls left to play hockey, leaving twice as many boys as girls in the classroom. What was the original number of students present?

(A) 24   (B) 32   (C) 48   (D) 54

382. Simplify: $\dfrac{x-2}{x^2-7x+6} \div \dfrac{x+6}{x^2+x-2}$

(A) $\dfrac{x^2-4x-12}{x^3-5x^2-8x+12}$ (B) $\dfrac{1}{9}$ (C) 1   (D) $\dfrac{x^2-4}{x^2-36}$ (E) $\dfrac{x^2-x-1}{x^2+3x-18}$

383. $(-1)^1+(-1)^2+(-1)^3+\ldots\ldots\ldots+(-1)^{98}+(-1)^{99}=$

(A) 1   (B) 0   (C) -1 (D) -99

384. A cube has edges of length 3. If P and Q are points on its surface, what is the maximum straight-line distance from P to Q?

(A) $3\sqrt{2}$ (B) $3\sqrt{3}$ (C) 6   (D) $3\sqrt{5}$

385. Determine the equation of the line that passes through the point $(8, -3)$ and is perpendicular to the line whose equation is $2x-3y-10=0$.

(A) $y=-\dfrac{3}{2}x+9$ (B) $y=-\dfrac{2}{3}x+\dfrac{7}{3}$ (C) $y=-\dfrac{3}{2}x-15$

(D) $y=-\dfrac{1}{2}x+1$ (E) $y=\dfrac{3}{2}x-15$

386. Solve the system $ax+by=3$ for x and y. The answers are expressed as (x, y).
$3ax+by=7$

(A) $(\frac{2}{a}, \frac{3}{b})$ (B) $(\frac{a}{2}, \frac{b}{3})$ (C) $(-\frac{2}{a}, \frac{3}{b})$ (D) $(\frac{2}{a}, \frac{1}{b})$

387. Solve for x: $\dfrac{x+1}{x-3} \geq 0$

(A) $-1 \leq x < 3$ (B) $x \leq -1$ or $x > 3$ (C) $-3 \leq x < 1$ (D) $x \leq -3$ or $x \geq 1$

388. $(x-y-z)^2 =$

(A) $x^2 + y^2 + z^2 - 2xy - 2xz + 2yz$ (B) $x^2 - y^2 - z^2 - 2xy - 2xz + 2yz$

(C) $x^2 + y^2 + z^2$ (D) $x^2 + y^2 + z^2 - 2xyz$

389. Your school cafeteria makes its delicious tuna salad by adding 2 pounds of mayonnaise to every 3 pounds of canned tuna. Canned tuna cost \$1.50 per pound and mayonnaise costs \$0.75 per pound. How many pounds of tuna salad can the cooks prepare for \$100?

(A) $83\dfrac{1}{3}$ (B) $33\dfrac{1}{3}$ (C) 50   (D) 75

390. If a is any positive integer, then which of the following is not a true statement?

(A) $2a+1$ is always an odd integer.   (B) $\sqrt{a}$ is always a real number.

(C) $\sqrt{-a}$ is always an imaginary integer   (D) a^3 is always an odd integer

391. If $\dfrac{n}{x^2 - 36} = \dfrac{1}{x-6} + \dfrac{1}{x+6}$, then $n =$

(A) x (B) $2x$ (C) $2(x+6)$ (D) $2(x-6)$

392. If $f(x) = x^3$ and $h(x)$ is obtained by shifting $f(x)$ down 4 units and right 2 units, then $h(x) =$

(A) $(x-2)^3 + 4$ (B) $(x+2)^3 - 4$ (C) $(x-2)^3 - 4$ (D) $(x-4)^3 - 2$

393. If the lines $y_1 = (n+2)x + 10$ and $y_2 = (n-4)x + 2$ are perpendicular, then n could equal which of the following?

(A) 3.45   (B) 3.83   (C) 1   (D) 0.5

395. The graph of $y = -x^4 + 12x - 18$

(A) intersect the x-axis at exactly one point.
(B) intersect the x-axis at exactly two point.
(C) intersect the x-axis at exactly four point.
(D) does not intersect the x-axis.

396. Write the following without radicals in the denominator: $\dfrac{\sqrt{5}}{\sqrt{5}+2}$

(A) $5 - 2\sqrt{5}$ (B) $5 + 2\sqrt{5}$ (C) $\dfrac{1}{2}$ (D) $\dfrac{5}{2}$

397. If $\dfrac{2}{x} + \dfrac{3}{y} = 21$ and $\dfrac{4}{x} - \dfrac{1}{y} = 7$, then $\dfrac{y}{x}$ equals

(A) 6   (B) $\dfrac{7}{3}$ (C) $-\dfrac{7}{3}$ (D) $\dfrac{3}{5}$

398. Find the equation of the line that perpendicular to the line $x - y = 4$ and passes through the point $(3, 2)$.

(A) $x - y = 1$ (B) $x + y = 5$ (C) $2x - y = 4$ (D) $x - y = -4$

399. $(\sec\theta+\tan\theta)(\sec\theta-\tan\theta)=$

(A) $\sec^2\theta$ (B) 1   (C) $\tan^2\theta-\sec^2\theta$ (D) -1

400. Find all the real solutions to the equation $\sqrt{6}\,x^2+2x-\sqrt{\dfrac{3}{2}}=0$.

(A) $-\dfrac{\sqrt{6}}{6},\dfrac{\sqrt{6}}{2}$ (B) $\dfrac{\sqrt{6}}{6},-\dfrac{\sqrt{6}}{2}$ (C) $\dfrac{-1\pm\sqrt{3}}{\sqrt{6}}$ (D) $\dfrac{-1\pm\sqrt{2}}{\sqrt{6}}$

401. In which quadrant(s) do the graphs of the equations $3x-5y+2=0$ and $y=0.6x+0.4$ intersect?

(A) I   (B) II   (C) III   (D) None of these

402. The sum of the solutions of $|x-3|=2|x+1|$ is

(A) $-\dfrac{14}{3}$ (B) $\dfrac{14}{3}$ (C) -5 (D) 5

403. The linear regression model $C=13.2m+20.5$ relates the calories burned using a new exercise machine (C) to the number of minutes a person uses the machine (m). Which of the following statements about this model must be true?

  I. When a person spends 22 minutes using the machine, the predicted number of calories burned is approximately 311.

  II. There is relatively no correlation of between C and m.

  III. A person burns approximately 13.2 calories each minute he or she uses the machine.

(A) I only   (B) III only   (C) I and II only   (D) I and III only

404. If (x, y) is a point on the graph of $f(x)$, then which of the following is a point on the graph of $f^{-1}(x)$?

(A) $(-x, -y)$ (B) $(-x, y)$ (C) (y, x) (D) $(x, -y)$

405. To earn an A in his math class, Daniel must correctly answer 80% of the questions on three regular tests plus the final exam, which count twice as much as a regular test. If Daniel has scores of 72, 67, and 75 on the tests which are worth 100 points, what percentage of total points does he need on the final exam to earn an A?

(A) 87%   (B) 90%   (C) 92%   (D) 93%

406. The domain of the function $f(x) = \dfrac{x+1}{x} - \dfrac{3}{\sqrt{5-x}}$ is

(A) $x \neq 0$ (B) $-\infty < x < \infty$ (C) $x > 5$ and $x \neq 0$ (D) $x < 5$ and $x \neq 0$

407. If $x_0 = 5$ and $x_{n+1} = x_n \sqrt{x_n + 4}$, then $x_4 =$

(A) 65.4   (B) 544.6   (C) 12,709.8   (D) 12,756.4

408. The best description of the line in the xy-plane that satisfies $x = 3$ is:

(A) A point at $(3, 0)$.   (B) A line that has a slope of 3.
(C) A line that is parallel to $y = 3x$.   (D) A line that is parallel to the y-axis.

409. Solve for y in terms of x if $x = \dfrac{2y-1}{y-5}$.

(A) $y = \dfrac{2x-1}{x-5}$ (B) $y = \dfrac{\frac{1}{2}x+1}{x+5}$ (C) $y = \dfrac{x-5}{2x-1}$ (D) $y = \dfrac{5x-1}{x-2}$

410. Solve for x: $\quad x+1=\dfrac{20}{x}$, $x \neq 0$

(A) 4 only   (B) -5 only   (C) -4 or 5   (D) -5 or 4

411. If the system $\begin{cases} y=x^2-x+9 \\ y=kx \end{cases}$ has exactly one solution, then $k=$

(A) -7 (B) 5   (C) -5 or 7   (D) 5 or -7

412. Simplify: $\dfrac{x^2}{x^{10}}$, $x \neq 0$

(A) $x^{\frac{1}{8}}$ (B) $x^{\frac{1}{5}}$ (C) x^{-5} (D) x^{-8}

413. If $\begin{cases} 3x-2y=12 \\ 5x+4y=-2 \end{cases}$ is a system of simultaneous equations, then $x=$

(A) -3 (B) -2 (C) 2   (D) 3

414. Find the least common denominator of $\dfrac{-5}{18x^2y^3}$ and $\dfrac{11}{12xy^2z}$.

(A) $6xyz$ (B) $36x^2y^3z$ (C) $36xy^2z$ (D) $36xyz$

415. How long is the base of an isosceles triangle if the other two sides measure 20 cm and each base angle $28°$?

(A) 12.7   (B) 28.3   (C) 30.6   (D) 35.3

416. A committee of 4 people is to be selected from a group of 8 women and 4 men. Assuming the selection is made randomly, what is the probability that the committee consists of 2 women and 2 men?

(A) $\dfrac{56}{165}$ (B) $\dfrac{6}{495}$ (C) $\dfrac{28}{495}$ (D) $\dfrac{28}{165}$

417. If f is a function that is the line with coordinates $(0,0)$ and $(2,14)$, then $f^{-1}(6.2) =$

(A) 0.44   (B) 0.76   (C) 0.89   (D) 1.11

418. If $f(x) = \sqrt{x+2}$ and $g(x) = x^2 - 2$ then $f(g(x)) =$

(A) x (B) $|x|$ (C) 2   (D) $\sqrt{x^2 - 2}$

419. If a, b, c are positive integers and $\dfrac{16}{5} = a + \dfrac{1}{b + \dfrac{1}{c}}$, then $a+b+c$ equals

(A) 5   (B) 8   (C) 10   (D) 12

420. The area of a rectangle is $6x^4 + 3x^3 - 4x^2 + 8x + 5$ and its width is $2x+1$. The length of the rectangle is

(A) $3x^3 + x^2 - 2x + 5$ (B) $3x^3 - 2x + 5$ (C) $3x^3 + 5$ (D) $3x^3 + x^2 + 2x + 5$

421. If $ab > 0$, then the two equations $x + y = a$ and $\dfrac{1}{x} + \dfrac{1}{y} = b$ have a unique simultaneous solution for x and y provided ab equals

(A) 1   (B) 2   (C) 4   (D) 8

422. If $\dfrac{1-x}{x}=\dfrac{4-4x}{x}$, then $\dfrac{1-x}{x}=$

(A) -1 (B) 0   (C) $\dfrac{1}{4}$ (D) 1

423. A number k is increased by 10. If the fifth root of that result equals -2, $k=$

(A) -42 (B) -22 (C) -11.1 (D) 22

424. If $\sin x = 0.45$, then $\cos\left(\dfrac{x}{2}\right)=$

(A) 0.45   (B) 0.52   (C) 0.71   (D) 0.97

425. Simplify the following rational expression: $\dfrac{a^3-b^3}{b^2-a^2}$

(A) $a-b$ (B) $b-a$ (C) $\dfrac{a^2+ab+b^2}{a+b}$ (D) $-\dfrac{a^2+ab+b^2}{a+b}$

426. A train goes from town A to town B. If it averages 50 mph (miles per hour), then will be 20 minutes late, and it averages 80 mph, then it will be 10 minutes early. If it goes 60 mph it will be

(A) early by 3 minutes   (B) early by $\dfrac{7}{3}$ minutes   (C) early by 5 minutes

(D) late by $\dfrac{20}{3}$ minutes

427. The product of the solutions of $5\sqrt{x+4}=x+10$ is

(A) -5 (B) 0   (C) 5   (D) 10

428. Suppose $a * b = ab - b^2$. Then $2 * (3 * 4) =$

(A) -28　　　(B) -24　　　(C) 2　　　(D) 24

429. What is the domain of $f(x) = \sqrt{9 - x^2}$?

(A) $x \leq 3$　　　(B) $x \geq -3$　　　(C) $-3 \leq x \leq 3$　　　(D) $x \leq -3$ or $x \geq 3$

430. If $7a^4 = 28a^2$, then which of the following are the possible values of a?

(A) 2　　　(B) -2 or 2　　　(C) 4　　　(D) -2 or 0 or 2

431. $f(x) = 2x + \sqrt[3]{x}$, then $f(f(8)) =$

(A) 18　　　(B) 36　　　(C) 38.6　　　(D) 40.2

432. How many 4-person committee can be formed from a group of 9?

(A) 24　　　(B) 126　　　(C) 1512　　　(D) 3024

433. If $2\sqrt{x} - \sqrt{2x+1} - 1 = 0$, then what is the value of x?

(A) -2　　　(B) 1　　　(C) 2　　　(D) 4

434. If the graph of the equation $y = mx + 4$ has points in the 4th quadrant, then which of the following must be true for m?

(A) $m = 0$　　　(B) $m < 0$　　　(C) $0 < m < 1$　　　(D) $m > 0$

435. A teacher gives a test to two algebra classes. The first class has 18 students, and the second has 24 students. If the average for the first class is 84%, and x is the overall average grade for all of the students, then which of the following statements is true?

(A) $84 \leq x \leq 93.1$ (B) $42 \leq x \leq 92$ (C) $36 \leq x \leq 93.1$ (D) $36 \leq x \leq 92$

436. Which of the following is a zero of $f(x) = 6x^3 - 5x^2 + 4x - 15$?

(A) 0.33   (B) 1.00   (C) 1.25   (D) 1.50

437. A triangle has sides measuring 4, 4, and 6 inches. What is the measure of its largest angle?

(A) $82.8\,°$ (B) $97.2\,°$ (C) $41.4\,°$ (D) $120.0\,°$

438. Which of the following is the equation of the circle whose diameter is the line segment connecting points $(1, -4)$ and $(3, 6)$?

(A) $(x-2)^2 + (y-1)^2 = 26$ (B) $(x-1)^2 + (y+4)^2 = 104$

(C) $(x+2)^2 + (y+1)^2 = 26$ (D) $(x+2)^2 + (y+1)^2 = 25$

x	y	z
0	2	0
3	3	6
4	4	8
5	5	10

439. For every row of values above, $\dfrac{x^2}{y} = kz$, where k is a constant. What is the value of k?

(A) $\dfrac{1}{6}$ (B) $\dfrac{1}{3}$ (C) $\dfrac{1}{2}$ (D) $\dfrac{3}{2}$

440. If $p \neq 0$ and p is inversely proportional to q, which of the following is directly proportional to $\dfrac{1}{p^2}$?

(A) $-\dfrac{1}{q^2}$ (B) $\dfrac{1}{q^2}$ (C) $\dfrac{1}{q}$ (D) q^2

441. The cost of a piece of a certain type of lumber is directly proportional to its length. A piece of this lumber that is 16 feet long costs \$12.00. What is the cost, in dollars, of a piece of this lumber that is x yards long? (1 yard=3 feet)

(A) $x-2$ (B) $3x-2$ (C) $\dfrac{3}{4}x$ (D) $\dfrac{9}{4}y$

442. A right circular cylinder has a height of 12 and a radius of 3. If X and Y are two points on the surface of the cylinder, what is the maximum possible length of XY?

(A) $3\sqrt{17}$ (B) 6   (C) $6\sqrt{5}$ (D) $5\sqrt{6}$

443. Some number n is added to the three numbers 1, 13, and 61 to create the first three terms of a geometric sequence. What is the value of n?

(A) 1   (B) 2   (C) 3   (D) 5

444. If $f(x) = x^2 + 1$, then $f(f(4)) =$

(A) 17   (B) 256   (C) 34   (D) 290

445. The amount of interest earned on savings is directly proportional to the amount of money saved. If $104 interest is earned on $1,300, how much interest will be earned on $1,800 in the same period of time?

(A) $72   (B) $96   (C) $124   (D) $144

446. A cell phone provider charges $0.40 for the first minute of a call and $0.20 for each additional minute or each additional portion of a minute. The cost of a call, C, is given by the model $C=0.40+0.20\bar{t}$ where t is the number of minutes over the first minute and $\bar{t}$ is the least integer greater than or equal to t. What is the cost of a 12.25-minute call?

(A) $2.60   (B) $2.65   (C) $2.80   (D) $2.85

447. Let (2, 4) be on the graph of $f(x)$. On the graph $f(x+3)$, the point becomes

(A) (5, 4)   (B) (−1, 4)   (C) (2, 1)   (D) (2, 7)

448. What is the remainder when the polynomial $x^4-5x^2-10x-12$ is divided by $x+2$?

(A) 4   (B) −36   (C) −20   (D) 2

449. What is the maximum value of $f(x)=3-(x+1)^2$?

(A) −3   (B) −1   (C) 1   (D) 3

450. The function $f(x)$ has the value 0 if and only if x is a member of the set $\{-3, 0, 1\}$. For what values of x is $f(x-3)=0$?

(A) $\{-3, 0, 1\}$ (B) $\{0, 1\}$ (C) $\{0, 3, 4\}$ (D) $\{-6, 0, -2\}$

451. If $f(x)$ is a function for which $f(2+k)=f(2-k)$ and $f(-3)=0$, for which of the following is $f(x)$ also equal to zero?

(A) $x=3$ (B) $x=5$ (C) $x=-1$ (D) $x=7$

452. In the xy-plane, the graph of the function f is a line. If $f(2)=6$ and $f(6)=20$, what is the value of $f\left(\frac{1}{3}\right)$?

453. If $\sin\theta = \frac{1}{2}\cos\theta$, then what is the smallest positive value of θ?

(A) 26.57° (B) 30° (C) 53.14° (D) 60

454. What is the sum of the integers from 1 to 300?

(A) 9000   (B) 44850   (C) 45150   (D) 90000

455. Which of the following lines are asymptotes of the graph of $f(x)=\dfrac{3(x^2-9)}{x^2-4}$?

  I. $x=\pm2$ II. $x=3$ III. $y=3$

(A) I only   (B) II only   (C) I and II only   (D) I and III only

456. Find the maximum profit if the equation of profit is $p(x)=2+10x-x^2$, where $p(x)$ is the profit and x is the number of products sold.

457. A projectile is fired vertically upward with the initial speed of 96 feet per second. The height of the projectile is given by $h(t) = vt - 16t^2$, where v is the initial speed and t is the time in seconds. What will be the maximum height of the rocket reaches?

459. The graph of $f(x) = x^3$ is translated 6 units up, 2 units right, and reflected over the x-axis. If the resulting graph represent $g(x)$, then $g(-1) =$

(A) -21 (B) 5   (C) 14   (D) 21

460. Find k so that the equation $2x^2 + 8x + k = 0$ has one real number (double) root.

461. $(6\sin x)(3\sin x) - (9\cos x)(-2\cos x) =$

(A) 1   (B) -18 (C) 18   (D) -1

462. If $f(x) = \sqrt[3]{2x^3 - 5}$, then $f^{-1}(2.5) =$

(A) 1.74   (B) 2.18   (C) 2.37   (D) 2.97

463. An open box was formed from a rectangular sheet of metal by cutting a square of side 4 cm from each of the corners of the rectangular sheet and then folding up the edges. The length of the original sheet of metal was $3 cm$ less than twice its width, and the volume of the box was 532 cm^3. Find the dimensions of the sheet of metal.

464. In a geometry exercise, there is a 0.15 probability a protractor is in error of $1°$ or more. If 5 protractors are used, what is the probability that all of them are in error of $1°$ or more?

(A) 0.000076   (B) 0.60   (C) 0.15   (D) 0.75

465. Find k so that the equation $4x^2 + 12kx + 9 = 0$ has one real number (double) root.

466. If a coin is tossed three times, what is the probability that exactly two heads will appear?

(A) $\dfrac{3}{8}$ (B) $\dfrac{3}{7}$ (C) $\dfrac{1}{2}$ (D) $\dfrac{1}{4}$

467. If a rectangular prism has faces with areas of 8, 10, and 20 units^2, then what is the volume?

(A) 20   (B) 40   (C) 80   (D) 400

468. Find all solutions of $\dfrac{3x^2}{x+5} = \dfrac{x^2 - 9x + 5}{x+5}$.

(A) -5 (B) $-\dfrac{1}{2}$ (C) $\dfrac{1}{2}$ (D) -5 and $\dfrac{1}{2}$

469. Find the slope of the line $2x - 5y = 10$.

(A) $-\dfrac{2}{5}$ (B) $\dfrac{5}{2}$ (C) $\dfrac{2}{5}$ (D) -2

470. If a function is an odd function, then $f(-x) = -f(x)$ for all values of x in the domain. Which of the following is an odd function?

(A) $f(x) = \sin x$ (B) $f(x) = \cos x$ (C) $f(x) = x^2 - 10$ (D) $f(x) = 4^x$

471. If $p \star q = \dfrac{2p-q}{p-2q}$, find $6 \star 2$ in simplest form.

(A) 2   (B) 3   (C) 5   (D) 8

472. The area of a rectangle is 192 ft^2. If the width is $\dfrac{3}{4}$ of the length, what is the width of the rectangle?

(A) 12 ft (B) 16 ft (C) 24 ft (D) 144 ft

473. $2^0 - (\dfrac{1}{2})^{-2} + (4-1)^{-1} =$

(A) -5 (B) $-\dfrac{8}{3}$ (C) $-\dfrac{5}{3}$ (D) $\dfrac{13}{12}$

474. $(\sqrt[7]{\sqrt[10]{\sqrt{7}}})^{35} =$

(A) $\dfrac{1}{7}$ (B) $\sqrt{7}$ (C) 7   (D) $7\sqrt{7}$

475. Solve for x: $-9(x-3)+2x \geq 2(4-x)$

(A) $x \leq -1$ (B) $-1 \leq x$ (C) $x \leq \dfrac{19}{5}$ (D) $-\dfrac{59}{5} \leq x$ C

476. What is the y-intercept of the line tangent to the circle $x^2 + y^2 = 1$ at the point $(\dfrac{3}{5}, \dfrac{4}{5})$?

(A) 0.80   (B) 1   (C) 1.20   (D) 1.25

477. Let $x = y + 3$. What is the value of $(x - y)^3$?

(A) $8b^2$ (B) -27 (C) -9 (D) 27

478. The selling price of a coat is $91 plus the overhead plus the profit. If the overhead is 20% of the selling price and the profit is 10% of the selling price, what is the selling price?

(A) $70   (B) $104   (C) $117   (D) $130

479. Solve: $(x + 5)^{\frac{1}{2}} + 1 = x$

480. What is the remainder when $x^{10} - 1$ is divided by $x + 2$?

481. Solve for x and y: $\begin{cases} 5x + y = 2 \\ (x + 1)^2 = y + 7 \end{cases}$

482. Solve the following system for x: $\begin{aligned} 2x - 3y &= 5 \\ -\frac{x}{2} + \frac{5y}{2} &= \frac{1}{2} \end{aligned}$

(A) 1   (B) 2   (C) 3   (D) 4

483. Multiply: $(x - 5)(x + 2)(x + 5)$

(A) $x^3 - 27x - 50$ (B) $x^3 - 50$ (C) $x^3 + 25x^2 + 2x + 50$

(D) $x^3 + 2x^2 - 25x - 50$

484. When a scientist dives in salt water to a depth of 9 feet below the surface, the pressure due to the atmosphere and surrounding water is 18.7 pounds per square inch. As the scientist descends, the pressure increases linearly. At a depth of 14 feet, the pressure is 20.9 pounds per square inch. If the pressure increases at a constant rate as the scientist's depth below the surface increases, which of the following linear models best describes the pressure p in pounds per square inch at a depth of d feet below the surface?

(A) $p = 0.44d + 0.77$ (B) $p = 0.44d + 14.74$ (C) $p = 2.2d - 1.1$

(D) $p = 2.2d - 9.9$

485. At a restaurant the cost for a breakfast taco and a small glass of milk is $2.10. The cost for 2 tacos and 3 small glasses of milk is $5.15. Which pair of equations can be used to determine t, the cost of a taco, and m, the cost of a small glass of milk?

(A) $t + m = 2.10$ (B) $t + m = 2.10$ (C) $t + m = 2.10$

  $2t + 2m = 5.15$ $3t + 3m = 5.15$ $3t + 2m = 5.15$

(D) $t + m = 2.10$

  $2t + 3m = 5.15$

486. Sixty cookies were to be equally distributed to x campers. when 8 campers did not want the cookies, the other campers each received 2 more cookies. Which of the following equations could be used to find the number of camper x?

(A) $x^2 - 8x - 240 = 0$ (B) $x^2 - 8x + 240 = 0$ (C) $x^2 + 8x - 240 = 0$

(D) $x^2 + 8x + 240 = 0$

487. Pat can inspect a case of watches in 5 hours. James can inspect the same case of watches in 3 hours. After working alone 1 hour, Pat stops for lunch. After taking 40 minutes lunch break, Pat and James work together to inspect the remaining watches. How long do Pat and James work together to complete the job?

(A) 1 hour 30 min   (B) 1 hour 45 min   (C) 2 hours

(D) 2 hours 15 min

488. Which of the following is the distance between the vertices of $y = x^2 - 5$ and $y = -x^2 + 4$?

(A) 1   (B) 3   (C) 5   (D) 9   (E) 10

489. The diameter and height of a right circular cylinder are equal. If the volume of the cylinder is 2, what is the height of the cylinder?

(A) 1.37   (B) 1.08   (C) 0.86   (D) 0.80

490. If $0 < x < 1$, what is the x-intercept of $= y = 2x^2 - 5x + 2$?

491. The equation $ax^2 + 3x + c = 0$ has two distinct roots. Which of the following is (are) possible?

       I. $a = 1, c = 2$ II. $a = -1, c = -2$ III. $a = -1, c = 2$

(A) I only   (B) II only   (C) I and II only   (D) I, II, and III

492. The minimum value of $f(x) = x^2 + x + 4$ is

(A) 2   (B) -2 (C) 4   (D) $\dfrac{15}{4}$

493. You throw a basketball whose path can be modeled by $y = -4x^2 + 16x + 6$ where x represents time and y represents height of the basketball. Which of the following is the maximum height?

(A) 18   (B) 20   (C) 22   (D) 24

494. The maximum value of $f(x) = -x^2 - 8x + 10$ is

(A) -6 (B) 6   (C) 10   (D) 26

495. For some real number t, the first three terms of an arithmetic sequence are $2t$, $5t - 1$, and $6t + 2$. What is the numerical value of the fourth term?

(A) 8   (B) 10   (C) 16   (E) 19

496. The equation of the line that passes through the points $(2, -3)$ and $(4, 0)$ is

(A) $3x - 2y - 12 = 0$ (B) $3x + 2y - 12 = 0$ (C) $3x - 2y - 8 = 0$
(D) $3x - 2y = 0$

497. Two executives are driving separately to a conference. The first leaves at 2 pm, traveling at 60 mph. The second leaves at 2:30 pm. What is the speed of travel of the second executive, if he was passing the first one at 4:30 pm?

(A) 65 mph   (B) 70 mph   (C) 75 mph   (D) 80 mph

498. Determine the answer: $\sqrt{3ab} \cdot \sqrt{3a^2} \cdot \sqrt{3a^4b}$

(A) $3a^3\sqrt{3ab}$ (B) $3a^3b\sqrt{3ab}$ (C) $9a^3b\sqrt{3a}$ (D) $3a^3b\sqrt{3a}$

499. Which of the following are factors of $6x^3 + 29x^2 - 7x - 10$?

        I. $3x - 1$ II. $2x + 2$ III. $x + 5$

(A) I only   (B) II only   (C) III only   (D) I and III only

500. Solve: $\begin{cases} 2x - 3y = 8 \\ 3x + 2y = -1 \end{cases}$

501. Solve: $x^2 - 3 = x(2x + 5)$

502. Let $f(x) = x^2 + 2$. For $g(t) = 2t - 1$, what is $f(g(x))$?

503. Let $x = \dfrac{y+1}{y-3}$. Write y in terms of x.

504. Solve: $\sqrt{x+3} = \dfrac{x}{2} + 1$

505. The concentration of a substance in a solution is 2×10^{-5} milligrams per milliliter. How many milligrams are 3×10^8 in milliliters?

(A) 6×10^{-13} (B) 5×10^{-13} (C) 5×10^3 (D) 6×10^3

506. A jar contains 16 gum balls: 5 are yellow, 8 are red, and 3 are blue. Two gum balls are selected at random from the jar one at a time. If the first gum ball is red, what is the probability that the second gum ball will also be red?

(A) $\dfrac{7}{16}$ (B) $\dfrac{1}{2}$ (C) $\dfrac{8}{15}$ (D) $\dfrac{7}{15}$

507. If m and n are any real numbers such that $0 < m < 1 < n$, which of the following must be true of the value of $\dfrac{m}{n}$?

(A) $1 < \dfrac{m}{n} < n$ (B) $m < \dfrac{m}{n} < n$ (C) $m < \dfrac{m}{n} < 1$ (D) $0 < \dfrac{m}{n} < m$

508. A circle with center $(-5, 1)$ is tangent to the y-axis in the standard (x, y) coordinate plane. What is the radius of this circle?

(A) 5   (B) 1   (C) $\sqrt{6}$ (D) 4

YES CLASS

New SAT Math

ANSWER

New SAT Math Problems from College Board

1.   B

2.   B

3. $\frac{21}{4} < 9x - 3 < \frac{27}{5}$

4.   B

5.   C

6.   24

7.   N/A

8.   B

9.   C

10.   B

11.   C

12.   B

13.   B

14.   C

15.   D

16.   PART1 63

        PART2 7,272

17.   10

18.   A

19.   B

20.   5 or 7

21.   B

22.   C

23.   A

24.   57

25.   D

26.   B

New SAT Math Practice

1.	C
2.	C
3.	C
4.	A
5.	D
6.	B
7.	B
8.	B
9.	C
10.	B
11.	D
12.	A
13.	D
14.	D
15.	D
16.	C
17.	A
18.	D
19.	A
20.	D
21.	B
22.	A
23.	C
24.	C
25.	C
26.	A
27.	D
28.	D
29.	D
30.	D
31.	D
32.	A

33.	B
34.	D
35.	C
36.	D
37.	B
38.	D
39.	D
40.	D
41.	C
42.	A
43.	D
44.	C
45.	D
46.	C
47.	C
48.	A
49.	C
50.	A
51.	D
52.	B
53.	D
54.	D
55.	A
56.	B
57.	B
58.	B
59.	C
60.	A
61.	D
62.	A
63.	A
64.	D
65.	D

66.	D
67.	D
68.	A
69.	A
70.	A
71.	D
72.	B
73.	A
74.	A
75.	D
76.	D
77.	B
78.	D
79.	B
80.	A
81.	D
82.	D
83.	A
84.	A
85.	C
86.	C
87.	B
88.	D
89.	A
90.	C
91.	C
92.	C
93.	B
94.	D
95.	A
96.	B
97.	A
98.	C

99.	A
100.	C
101.	B
102.	C
103.	C
104.	A
105.	C
106.	D
107.	C
108.	B
109.	B
110.	10
111.	A
112.	B
113.	5 or 7
114.	A
115.	B
116.	C
117.	C
118.	D
119.	A
120.	C
121.	C
122.	A
123.	B
124.	C
125.	D
126.	5/2 sec
127.	L:12 W:6
128.	D
129.	A
130.	D
131.	D

132.	D
133.	B
134.	A
135.	D
136.	A
137.	B
138.	C
139.	A
140.	D
141.	B
142.	N/A
143.	A
144.	B
145.	C
146.	D
147.	$-4 < x < 4$
148.	C
149.	A
150.	C
151.	A
152.	B
153.	C
154.	B
155.	$\frac{21}{4} < 9x - 3 < \frac{27}{5}$
156.	B
157.	D
158.	D
159.	A
160.	C
161.	C
162.	24
163.	C
164.	D

165. C

166. D

167. A

168. D

169. D

170. D

171. C

172. D

173. D

174. C

175. B

176. D

177. B

178. C

179. D

180. C

181. B

182. A

183. E

184. C

185. A

186. B

187. B

188. B

189. C

190. 16

191. D

192. A

193. D

194. B

195. C

196. B

197. B

198.	E
199.	18
200.	A
201.	A
202.	B
203.	D
204.	75/2
205.	D
206.	C
207.	D
208.	A
209.	D
210.	D
211.	B
212.	B
213.	C
214.	D
215.	A
216.	B
217.	C
218.	A
219.	C
220.	D
221.	D
222.	D
223.	C
224.	B
225.	A
226.	C
227.	D
228.	C
229.	C
230.	C

231.	D
232.	B
233.	A=−3, B=−7/6, C=181/12
234.	A
235.	D
236.	B
237.	B
238.	A
239.	D
240.	x>5 or x<−2
241.	D
242.	B
243.	A
244.	C
245.	C
246.	A
247.	A
248.	A
249.	D
250.	C
251.	C
252.	B
253.	A
254.	D
255.	D
256.	D
257.	D
258.	D
259.	C
260.	B
261.	C
262.	A
263.	B

264.	B
265.	B
266.	N/A
267.	D
268.	D
269.	C
270.	A
271.	B
272.	D
273.	A
274.	A
275.	D
276.	B
277.	D
278.	C
279.	B
280.	C
281.	D
282.	E
283.	D
284.	A
285.	D
286.	B
287.	D
288.	B
289.	B
290.	C
291.	A
292.	C
293.	D
294.	D
295.	D
296.	C

297.	C
298.	A
299.	A
300.	A
301.	C
302.	D
303.	D
304.	C
305.	A
306.	B
307.	C
308.	D
309.	D
310.	D
311.	C
312.	A
313.	D
314.	D
315.	C
316.	D
317.	N/A
318.	D
319.	A
320.	C
321.	D
322.	A
323.	B
324.	D
325.	B
326.	D
327.	A
328.	B
329.	B

330.	D
331.	B
332.	D
333.	D
334.	D
335.	D
336.	D
337.	C
338.	D
339.	B
340.	A
341.	C
342.	B
343.	D
344.	B
345.	A
346.	A
347.	B
348.	B
349.	A
350.	C
351.	A
352.	C
353.	C
354.	A
355.	D
356.	D
357.	A
358.	B
359.	C
360.	D
361.	D
362.	B

363.	C
364.	D
365.	A
366.	A
367.	A
368.	C
369.	C
370.	B
371.	B
372.	A
373.	C
374.	D
375.	C
376.	C
377.	B
378.	D
379.	A
380.	D
381.	B
382.	D
383.	C
384.	B
385.	A
386.	D
387.	B
388.	A
389.	A
390.	D
391.	B
392.	C
393.	D
394.	N/A
395.	D

396.	A
397.	D
398.	B
399.	B
400.	B
401.	D
402.	A
403.	D
404.	C
405.	D
406.	D
407.	D
408.	D
409.	D
410.	D
411.	D
412.	D
413.	C
414.	B
415.	D
416.	A
417.	C
418.	B
419.	B
420.	B
421.	A
422.	B
423.	A
424.	D
425.	D
426.	D
427.	B
428.	B

429.	C
430.	D
431.	C
432.	B
433.	D
434.	B
435.	C
436.	D
437.	B
438.	A
439.	C
440.	D
441.	D
442.	C
443.	C
444.	D
445.	D
446.	D
447.	B
448.	A
449.	D
450.	C
451.	D
452.	3
453.	A
454.	C
455.	D
456.	27
457.	144 ft
458.	N/A
459.	D
460.	8
461.	C

462.	B
463.	27 * 15
464.	A
465.	k=1 or −1
466.	A
467.	B
468.	C
469.	C
470.	A
471.	C
472.	A
473.	B
474.	B
475.	C
476.	B
477.	D
478.	D
479.	$\frac{1 \pm \sqrt{17}}{2}$
480.	1023
481.	(−8, 42), (1, −3)
482.	D
483.	D
484.	B
485.	D
486.	A
487.	A
488.	D
489.	A
490.	1/2
491.	D
492.	D
493.	C
494.	D

495.   D

496.   A

497.   C

498.   D

499.   C

500. $(1, -2)$

501. $\frac{-5 \pm \sqrt{13}}{2}$

502. $4x^2 - 4x + 3$

503. $y = \frac{3x+1}{x-1}$

504. $x = \pm 2\sqrt{2}$

505.   D

506.   A

507.   D

508.   A